Praise for *The Brain in Love*

"Amen tackles the neuroscience of attraction like a tour guide in a candy store." —*Los Angeles Times*

"A scientist, a lover, and a parent, Dr. Amen understands who we are as couples, and helps us to become the best lovers and friends we can be."

—MICHAEL GURIAN, author of *What Could He Be Thinking?: How a Man's Mind Really Works*

"Dr. Amen does it again! He offers a brilliant explanation of how the state of our brain affects our relationships and sexuality. He goes on to give excellent advice on how to understand your own and your partner's (or potential partner's) behavior, how to avoid the pitfalls in a relationship, and how to create the life you really want!"

—HYLA CASS, M.D., author of *8 Weeks to Vibrant Health*

Praise for *Making a Good Brain Great*

Winner of Audiophile's Earphone Award

"This book is wonderful. It gives the reader great understanding and hope that changes in oneself can be made. If you put these changes into action, a happy and healthy brain is yours."

—BILL COSBY

"Each of Daniel Amen's books contains special nuggets that can be found nowhere else, as well as a wealth of useful, general information that he brings together under one roof. This book offers excellent advice as well as a great deal of new information. An extremely useful and easy-to-read book."

—EDWARD HALLOWELL, M.D., author of *Delivered from Distraction* and *The Childhood Roots of Adult Happiness*

Praise for *Change Your Brain, Change Your Life*

New York Times Bestseller

"Cutting-edge technology, clinical wisdom, and heartfelt guidance come together to create a 'user's manual for enhancing the human brain.' Dr. Amen offers simple, direct, and immediately applicable prescriptions that can change anyone's life."

—EMMETT E. MILLER, M.D., author of
Deep Healing: The Essence of Mind/Body Medicine

"Dr. Amen's groundbreaking work will forever change the fields of psychiatry and psychology. A healthy brain is prerequisite to a healthy life. Dr. Amen provides a practical guide."

—EARL HENSLIN, PH.D., author of *You Are Your Father's Daughter* and *Man to Man: Helping Fathers Relate to Sons and Sons Relate to Fathers*

"This book is brilliant. Dr. Amen is a pioneer."

—JONATHAN WALKER, M.D., psychiatrist

12 LESSONS TO ENHANCE YOUR LOVE LIFE

THREE RIVERS PRESS • NEW YORK

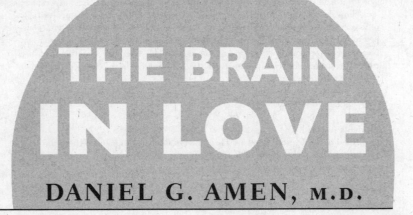

THE BRAIN
IN LOVE

DANIEL G. AMEN, M.D.

Previously published as *Sex on the Brain*

Published in the United States by Three Rivers Press, an imprint of the Crown Publishing Group, a division of Random House, Inc., New York.
www.crownpublishing.com

Three Rivers Press and the Tugboat design are registered trademarks of Random House, Inc.

Originally published in hardcover in the United States as *Sex on the Brain* by Harmony Books, an imprint of the Crown Publishing Group, a division of Random House, Inc., New York, in 2007.

The Library of Congress has cataloged *Sex on the Brain* as follows:

Amen, Daniel G.
    Sex on the brain : 12 lessons to enhance your love life / Daniel G. Amen.
    Includes bibliographical references and index.
  1. Sex (Psychology)—Health aspects.   2. Sex instruction.   3. Psychosexual disorders.   4. Brain.   I. Title.
    RA488.A46 2007
    155.3'1—dc22   2006021218

ISBN 978-0-307-58789-3

Printed in the United States of America

Design by Helene Berinsky

10 9 8 7 6 5 4 3 2 1

Contents

THE BRAIN
IN LOVE

The Brain Is the Largest Sex Organ (and Size Matters!)

- As you walk by me, millions of nerve cells spark in my brain and I have to turn to look at you again.
- You look back at me and a soft, brief smile forms on your lips. As you notice my eyes following you, your smile triggers an adrenaline release that causes my heart to leap with excitement.
- Chemicals send increased blood flow to sensitive areas, as thoughts of you light the emotional fire centers of my mind.
- For a brief moment you literally live in my skin.
- As we connect, my mind works overtime obsessing on your smell and the color of your eyes.
- You beat in my heart.
- You pulse in my nervous system from the nerve pathways of my brain to the soles of my feet.
- I start to become disoriented when we are apart.
- Over time, your touch becomes essential.
- I crave you.
- Your body feels warm and reassuring.
- I need it next to me.
- I sleep peacefully knowing you are near, and wake often in the dark to feel your skin.

- I never want to get out of bed when I am lying next to you.
- I look for you in my brain when you are away.
- Your voice sweetens the vibrations in the air.
- My mind beseeches me to make love to you, again and again.
- Our bodies navigate space together.
- Your mind reads mine as you know how I want to be touched.
- How does that happen?
- You must have cells that mirror my desires.
- The neurons of my eyes light up with sparks when you walk in a room, especially if you have been away for a while.
- Songs, smells, places, and pictures never let me forget you as they trigger the memory centers in my brain where you live as if you were next to me.
- The judgment part of my brain watches what I say when we are together so I can protect your feelings.
- I watch how your eyes, face, and body move as you talk to me, to know if you are happy, desirous, or in need of a hug or understanding.

Even though it feels genital, the vast majority of love and sex occurs in the brain. Your brain decides who is attractive to you, how to get a date, how well you do on the date, what to do with the feelings that develop, how long those feelings last, when to commit, and how well you do as a partner and a parent. Your brain helps you be enthusiastic in the bedroom or drains you of desire and passion. Your brain helps you process and learn from a breakup or makes you vulnerable to depression or obsession. When the brain works right, it helps you be thoughtful, playful, romantic, intimate, committed, and loving with your partner. When the brain is dysfunctional, it causes you to be impulsive, distracted, addicted, unfaithful, angry, and even hateful, thus ruining chances for continued intimacy and love.

Your brain is also the seat of orgasms. Some research implicates the right hemisphere of the brain. In fact, certain forms of epilepsy, especially those found in the right temporal lobe, have been associated with spontaneous orgasms. In one case from Taiwan, a forty-one-year-old woman had seizures that were induced only when she brushed her teeth. The seizure started with the feelings of being sexually aroused, then she felt an orgasmlike euphoria wash over her, which was followed by feelings of confusion. Her brain-imaging studies showed problems in the right temporal lobe, an area that has been associated with both orgasms and religious experience. When someone has orgasmiclike feelings when brushing her teeth, odds are that she will have very clean pearly whites.

Scientists agree that the brain is the organ of behavior; as such, it really is the largest sex organ in the body (about three pounds), and in this case size really does matter. Our brain becomes less and less active and decreases in overall size as we age. This is true for males and females and there appears to be an equal loss of gray matter (nerve cell bodies) and white matter (the connections between nerve cells). If you learn to take care of your brain, however, it can be active and healthy throughout your life. With targeted interventions, you can impact brain health, lose less brain tissue, and keep your brain healthy well into your elderly years.

Why does this matter to sexual function? As the brain dims in activity over the decades, so, too, does many people's sexual function. The two go together. In men between the ages of forty and seventy studied over a nine-year period, there was a significant decline in sexual function with age. This is consistent with past studies that have shown a decline in sexual desire, intercourse, and erection frequency. Erectile dysfunction (ED) is very common and increases with age. Forty percent of men in their forties, and 70 percent in their seventies had problems. In women, aging and menopause often negatively affect sexual interest and performance.

A major reason underlying both sexual and brain dysfunction is decreased blood flow. Blood does so many important things. It brings oxygen, sugar, and nutrients to your cells and it takes away waste products. Anything that interferes with healthy blood flow will impair an organ's functioning. Decreased blood flow to genitals from hypertension; vascular disease; diabetes; toxic exposure, such as drug abuse or smoking; physical trauma; and other causes impairs sexual function. Increased blood flow, from targeted interventions including exercise, ginkgo, and compounds that increase nitric oxide, such as Viagra and ginseng, improves function and reverses aging.

Likewise, decreased blood flow to your brain, from any cause, decreases brain function, which means you are likely to make impaired decisions and subsequently have less sex. Few scientists have looked at the connection between brain health and sexual behavior. That's where I come in. My primary work is as a brain-imaging specialist. I have been doing imaging work for more than sixteen years and my clinics have the world's largest database of scans related to behavior, more than 35,000. We look at the brain on a daily basis using a sophisticated study called SPECT imaging. SPECT stands for "single photon emission computed tomography," a nuclear medicine study that evaluates blood flow and activity patterns in the brain. We have looked at many healthy brains and brains in trouble. We have looked at the brains of children, teenagers, adults, and the elderly. We have looked at brains on medications, drug and alcohol abuse, supplements, prayer and meditation, gratitude, and a wide variety of psychological and biological treatments. We have looked at the brain in love, lust, commitment, divorce, domestic violence, sexual abuse, and loss.

At our clinics, our primary work is to help maximize people's brain function for the most satisfying and healthy life possible. We help healthy people who want to improve their own brain function, as well as treat attention deficit disorders (ADD), mood and anxiety disorders, obsessive-compulsive disorders, addictions, temper problems, and memory disturbances. We often help individuals and couples who struggle with relationship and sexual problems of all kinds. It is really possible to dramatically improve brain function, whether your brain is troubled or not, and thus dramatically improve your life. Our guiding principle for the past sixteen years has been "Change your brain, change your life."

Since most people cannot see the brain, it is often left out of the equations of our lives. Yet, it is at the core of our personal universe. Connecting sex and the brain through the lens of brain imaging has been one of the most fascinating journeys of my life, and I will share it with you in this book. I became much more effective in helping couples when I started looking at their relationships and sexuality together with brain function. It is clear that healthy brain function is associated with more loving and sexual relationships, while poor brain function is associated with more fighting, less sex, and higher divorce rates. In committed relationships, sex is a critical ingredient for health and longevity, but most people never connect the brain and sex.

I start with a clear bias: Sex is best in the context of a committed, loving relationship. Anthropologist Helen Fisher writes, "Do not copulate with people you do not want to fall in love with, because you might do just that." Sex bonds you to others, and in some cases, if you are not careful and thoughtful, it can put you in bondage to others. Although this is my bias, it is not always the context of some of the research studies I will share with you on the sexual benefits for health and longevity, which are based solely on sexual frequency. Having acknowledged that fact, there are other studies that strongly suggest a happy marriage is also associated with longevity, which usually means not sharing yourself sexually outside your primary relationship. The discussion throughout the book is on heterosexual relationships, but the same principles apply to all committed, loving relationships.

Based on my latest research, this book will share twelve practical neuroscience lessons to enhance your love and sex life. *Practical neuroscience* is a term I coined for the study of applying the latest brain research to everyday life. I am the type of person, like many of you, who always wants to know why I should learn something. If it isn't practical or helpful, then I don't want to expend great amounts of neuronal effort on it. The reason to study neuroscience is that it is immensely practical. Here's an example:

On a recent faculty retreat with the University of California, Irvine Department of Psychiatry, where I teach psychiatric residents, I was walking back from dinner through a shopping district along the quaint cobblestone streets of Taormina, Sicily, with one of my colleagues and his wife. They were talking about buying shoes. The wife wanted her husband to go with her to the shoe boutique and he was balking a bit. I looked at him and said, "You want to go with her." He gave me a quizzical look which said, "Why?"

I replied, "In the brain, the sensory area of the foot is right next door to the sensory area for the clitoris. Unknowingly, women often feel that buying shoes is like foreplay. Feet are one of the best ways to a woman's affections."

With my friend's help, his wife bought three pairs of shoes the next day. He had a smile on his face for the rest of the trip.

METHUSELAH'S SECRET

—

Sexuality, Longevity, Health, and Happiness

"Sex . . . what else is free, fun, low calorie, and exercise?"
—BARBARA WILSON, MD, neurologist and pain specialist

In August 1982, during my internship year on the sterile surgical floor at the Walter Reed Army Medical Center in Washington, D.C., Jesse was discharged from the hospital. He had been admitted for an emergency hernia operation two weeks earlier and there had been some minor complications. I remember Jesse so vividly now because he was one hundred years old, but talked and acted like a man thirty years younger. Mentally, he seemed every bit as sharp as any patient I had talked to that year or since. He and I developed a special bond, because unlike the surgery interns who spent a maximum of five minutes in his room each day, I spent hours over the course of his hospitalization talking to him about his life. The other interns were excited to learn about the latest operating techniques. I was interested in Jesse's story and I wanted to know about Jesse's secrets for longevity and happiness.

Jesse had his hundredth birthday in the hospital and it was quite an event. His wife, actually his second one, who was three decades younger, planned the event with the nursing staff. There was great love, playfulness, and physical affection between Jesse and his wife. Clearly, they still had the "hots" for each other.

Just before his discharge from the hospital, he saw me at the nurses' station writing notes. He enthusiastically waved me over to his room. His bags were packed and he was dressed in a brown suit, white shirt, and a blue beret. He looked deeply into my eyes as he quietly asked me, "How long, doc?"

"How long what?" I answered.

"How long before I can make love to my wife?"

I paused and he continued in a hushed voice, "You want to know the secret to live to a hundred, doc? Never miss an opportunity to make love to your wife. How long should I wait?"

A slow smile came over my face, "I think a week or so and you should be fine. Be gentle at first." Then I gave him a hug and said, "Thank you. You have given me hope for many years to come."

Science finally caught up to Jesse twenty-five years later. Now there is a wealth of research connecting healthy sexual activity to longevity. The lesson from Jesse still rings true today. While there are many ingredients to a long life—good genes, a positive outlook, a curious mind, and exercise—frequent sexual activity is one of them, too.

Like Jesse, Methuselah knew the secret, too. The oldest living Hebrew patriarch mentioned in the Bible, Methuselah was 187 years when his son Lamech was born. According to the author of the book of Genesis, he lived another 782 years, dying at the remarkable age of 969. Since then the name Methuselah has become a synonym for longevity; in this chapter I will illustrate what I consider Methuselah's Secret—the link between sexual frequency, sexual enjoyment, and longevity.

Can Ten Thousand Men Be Wrong?

Why do we have sex on the brain? Why is it the topic of so many conversations? From a scientific perspective, the answer is simple. More than any other basic need, sexual activity makes it possible for us to live beyond ourselves and for our species to survive. It is one of the most powerful drives motivating behavior. Yet, with the changing gender roles in our society and the constant mix of religious and cultural messages, sex has become confusing and frustrating. Is sex good? Is it bad? Is it important? Is it a luxury for pleasure or just a means of having children? New research shows that sex is important, not only to the physical survival of the human race, but to the survival of individuals as well. Sex is a natural part of being alive, of being human; it is healthy to want to express yourself in that way. Having a healthy, satisfying sex life is important for each person individually, but also to the health of romantic relationships.

Sexual interest, activity, and meaning change throughout the lifespan. Teenagers explore a new, exciting, confusing, and potentially risky (pregnancy and STDs) part of life. While young adults attempt to develop a sense of sexual competency, older people strive more for a sense of meaning in their sexual lives. Underlying most sexual contact is a desire for pleasure, release, and emotional connection.

Toward the end of Chuck's marriage, he was in bed reading Dean Ornish's book *Love and Survival*. In the book Dr. Ornish wrote about a study where ten thousand men were asked one question: "Does your wife show you her love?" The men who answered no, in significant numbers, died earlier. "Oh my God," Chuck thought, "I'm doomed." He had been in a twenty-year marriage where their libidos were badly matched, and he was turned away nine times out of ten. At 3 A.M. that morning he awoke with crushing chest pain. His heart checked out okay the next day, but he knew that something had to change. His life might depend on it.

Withholding sex, as a weapon of control or punishment, is common in relationships for both males and females. I have seen it in my practice for twenty-five years. Unfortunately, it is a deadly weapon and often kills relationships. After reading this chapter and understanding the research associated with frequent sexual activity and sexual enjoyment, my hope is that you will honor sex in your life. If you are someone who withholds sex as a way to punish your partner, my hope is that you will realize two things: One, the act of withholding physical affection is actually bad for you, as you miss out on its many benefits; and two, it puts your partner's health at risk. No kidding. I often joke in my lectures that if your partner knew the research, and you were withholding sex, he or she could potentially sue you for attempted murder. There is a lot of nervous laughter at this point. Of course, there are other reasons besides their partner withholding sex that people are not getting it, such as they are without a partner, there is an illness that affects sexual desire or performance, or people may be uncomfortable with their bodies.

Most of the research discussed here on the health benefits of sex involves sexual activity with a partner. Some of the research, however, has to do with orgasm frequency, which may also be due to masturbation. Sexual gratification or release through masturbation may be helpful for the brain. From a psychiatric standpoint, it is a complex issue. Masturbation can bring on a release of tension, but in some vulnerable people can also lead to excessive or addictive activity.

Healing: Sex Is the Best Medicine

Many studies have investigated the relationship between sexual activity and physical health. The potential dangers of sexual activity, including sexually transmitted diseases and unplanned pregnancies, have been widely reported, and rightly so. However, less publicized studies suggest that thoughtful sexual activity with a committed partner improves well-being by enhancing longevity, immune system function, joy, pain management, and sexual and reproductive health. These studies illustrate that sexual activity may be a preventive measure against the two leading causes of death in the United States, heart disease and cancer. Below are some of the aspects of your health that sex can improve.

Longevity

Learning how to enhance the largest sex organ in the body (the brain) and using it well to intimately connect with others may add years to your life and is likely to make you much happier. Serious research on sexuality began in the United States in the 1940s by Alfred Kinsey. He reported that sex reduces stress, and that people who have fulfilling sex lives are less anxious, less violent, and less hostile. Current research bears this out, as physical touch increases the hormone oxytocin, which boosts trust and lowers cortisol levels, the hormone of chronic stress. In a study done at Duke University, researchers followed 252 people over twenty-five years to determine the lifestyle factors important in influencing lifespan. Sexual frequency and past and present enjoyment of intercourse were three of the factors studied. For men, frequency of intercourse was a significant predictor of longevity. While frequency of intercourse was not predictive of longevity for women, those who reported past enjoyment of intercourse had greater longevity. This study suggested a positive association between sexual intercourse, pleasure, and longevity.

A 1976 report in *Psychosomatic Medicine* concluded that an inability to reach orgasm may have a negative impact on women's hearts. Only 24 percent of women in the healthy control group reported sexual dissatisfaction; while 65 percent of the women who had heart attacks reported trouble with sex. In this study, the two most common causes of dissatisfaction in women were due to impotence and premature ejaculation on the part of their husbands. Sexual health is not just an individual issue. It affects both parties' satisfaction and overall health.

A Swedish study found increased risk of death in men who gave up sexual intercourse earlier in life. The research was done on four hundred elderly men and women. At age seventy they were given a survey of their sexual activity and then followed over time. Five years later the death rates were significantly higher among the men who ceased sexual activity at earlier ages.

A daring group of researchers from Queen's University in Belfast, Ireland, included in a long-term study of health a question about sexual activity. The authors studied nearly one thousand men between the ages of forty-five to fifty-nine living in or near Caerphilly, Wales, and recorded the frequency of sexual intercourse each week and month. The researchers then divided the men into three groups: high orgasm frequency (those who had sex twice or more a week), an intermediate group, and low orgasm frequency (those who reported having sex less than monthly). The men were monitored again ten years later. Researchers found that the **death rate from all causes for the least sexually active men was twice as high as that of the most active group.** The death rate in the intermediate group was 1.6 times greater than for the active group.

Many questions come to mind with this type of study, such as "Is it the orgasm that is healing? Or, the touch and physical and emotional connection that comes with intercourse? Does poor health decrease sexual activity? Do other factors such as lack of exercise, alcohol, and depression cause both poor health and less sexual activity?" The researchers found that the robustness of their findings persisted even after adjusting for differences in age, social class, smoking, blood pressure, and evidence of existing coronary heart disease at the initial interview. This suggests a more likely protective role of sexual activity.

The Irish researchers wrote, "The association between frequency of orgasm and mortality in the present study is at least—if not more—convincing on epidemiological and biological grounds than many of the associations reported in other studies and deserves further investigation to the same extent. Intervention programs could also be considered, perhaps based on the exciting, 'At least five a day' campaign aimed at increasing fruit and vegetable consumption—although the numerical imperative may have to be adjusted."

In a 2001 follow-up study, this same research group found that having sex three or more times a week reduced by half the risk in males of having a heart attack or stroke. If a drug company came up with a medicine that performed as well, their stock would soar through the roof of Wall Street. The coauthor of the study, Shah Ebrahim, PhD, underscored the results by saying, "The relationship found between frequency of sexual intercourse and mortality is of considerable public interest." There is truth to the saying that an apple a day keeps the doctor away. It may also be true that an orgasm a day keeps the coroner away.

Fewer Sick Days

A study from the Institute for Advanced Study of Human Sexuality conducted by Dr. Ted McIlvenna looked at the sex lives of ninety thousand American adults. He found that sexually active people take fewer sick leaves and enjoy life more.

Boosted Immunity

According to gynecologist Dr. Dudley Chapman, orgasms boost infection-fighting cells up to 20 percent. Psychologists at Wilkes University in Pennsylvania found that students who had regular sexual activity had a third higher levels of immunoglobulin A (IgA), an antibody that boosts the immune system and can help fight colds and flu.

Healthy Sexual and Reproductive Behavior

Research done by Dr. Winnifred Cutler, a specialist in behavioral endocrinology, indicated that women who have intercourse with a male partner at least once a week are likely to have more regular menstrual cycles than women who are celibate or who have infrequent sex. In same-sex couples, women who engaged in sexual activity at least three times per week also had more regular cycles. In her "White Paper for Planned Parenthood," Dr. Cutler reported that sexual and reproductive health of both women and men is influenced by their sexual activity. She reports that regular sex can have positive effects also on reproductive health. Here are several examples:

Fertility. Frequent sexual activity may enhance fertility. Studies of menstrual cycle variability and frequency of intercourse have demonstrated that regular intimate sexual activity with a partner promotes fertility by regulating menstrual patterns.

Menstrual Cycle Regularity. A series of studies found that women who engaged in intercourse at least once per week had cycle lengths that were more regular than women who had sex sporadically or who were celibate.

Relief of Menstrual Cramps. Nine percent of nineteen hundred women stated that they masturbated in the previous three months to relieve menstrual cramps.

Pregnancy. A review of fifty-nine studies that were written between 1950 and 1996 concluded that sexual activity during pregnancy does not harm the fetus, as long as there are no other risk factors, such as sexually transmitted diseases, involved. In addition, some research has shown that sexual activity throughout pregnancy may serve as a protection against early delivery, especially during the third trimester (between the twenty-ninth and thirty-sixth weeks). Of more than eighteen hundred women, excluding those who could not have sex for medical reasons, preterm delivery was significantly reduced in the women who had intercourse late in their pregnancy.

Healthy Prostate. The prostate gland is responsible for producing some of the secretions in semen. Sometimes the prostate becomes inflamed and painful (prostatitis). In single men who had prostatitis, more than 30 percent who masturbated more frequently reported marked or moderate improvement of their symptoms. In addition, there is a suggestion that frequent ejaculation may help prevent chronic nonbacterial infections of the prostate.

Higher Youth Hormone Levels
(DHEA, Estrogen, and Testosterone)

Dr. Cutler also reported that women who enjoy regular sex had significantly higher levels of estrogen in their blood than women experiencing either infrequent sex or no sex at all. The benefits of estrogen include a healthy cardiovascular system, lower bad cholesterol, higher good cholesterol, increased bone density, and smoother skin. There is also growing evidence that estrogen is beneficial to brain functioning.

Another important hormone that seems to be affected by sexual activity is DHEA. Before orgasm the level of DHEA spikes in the body to several times higher than normal. DHEA is believed to improve brain function, balance the immune system, help maintain and repair tissue, promote healthy skin, and possibly improve cardiovascular health.

Testosterone is increased through regular sexual activity. Testosterone can help strengthen bones and muscles, and is also beneficial to a healthy heart and brain. The risk for Alzheimer's disease is twice as high for people with lower testosterone levels. Low testosterone levels are also associated with a low libido. From this connection one could infer that if you are not interested in sex, your memory may be in jeopardy as well.

Potential Cancer Prevention

A study conducted by Graham Giles from Australia concluded the more that men between the ages of twenty and fifty ejaculate, the less likely they are to develop prostate cancer. A study published by the *British Journal of Urology International* asserted that men in their twenties can reduce by a third their chance of getting prostate cancer by ejaculating more than five times a week.

Researchers have suggested that sexual expression may lead to a decreased risk of cancer because of the increase in levels of oxytocin and DHEA, which are associated with arousal and orgasm in women and men. A 1989 study found increased frequency of sexual activity was correlated with a reduced incidence of breast cancer among women who had never had a child. The study examined fifty-one French women who were diagnosed with breast cancer less than three months prior to the interview. They were matched with ninety-five controls. A higher risk of breast cancer also correlated with a lack of a sex partner and rare sexual intercourse, defined as less than once a month.

More Restful Sleep

Sexual release can help people go to sleep. Orgasm causes a surge in oxytocin and endorphins that may act as a sedative. One study found that 32 percent of 1,866 U.S. women who reported masturbating in the previous three months did so to help go to sleep. As most women know, men often go to sleep shortly after having sex.

Pain Relief

Studies have shown that orgasms can help treat some types of pain. Research by Beverly Whipple and Barry Komisaruk of Rutgers University found that through regular orgasm women had higher pain thresholds when suffering from conditions ranging from whiplash to arthritis. Immediately before orgasm, levels of the hormone oxytocin surge to five times their normal level. This in turn releases endorphins, which alleviate the pain. In women, sex also promotes the production of estrogen, which can reduce the pain of PMS.

Dr. Whipple's research identified the female G-spot, the vaginal "on switch" for female arousal, on the front inside wall of the vagina, opposite the clitoris. She showed that gentle pressure to this area raised pain thresholds by 40 percent and that during orgasm women could tolerate up to 110 percent more pain. In brain-imaging research to understand this finding, Dr. Whipple found that during peak arousal, the pain-killing center deep in the brain is activated. Signals from this part of the brain give orders to the body to release endorphins and corticosteroids. These chemicals help to temporarily numb the pain from many different causes. Activating this region also has a calming effect and can reduce anxiety. See Lesson Nine for more details on the G-spot.

Migraine Relief

Research suggests that when your partner says, "Not tonight, honey, I have a headache," you can help her with a loving roll in the sack. A study from the Southern Illinois University School of Medicine found that having an orgasm could help alleviate the pain from migraines. Among fifty-two migraine sufferers, sixteen reported considerable relief after an orgasm and another eight had their headaches completely gone. Since 2001, a couple of case studies reported that orgasm did help with pain relief. An earlier study of eighty-three women who suffered from migraine headaches showed that orgasm resulted in at least some relief for more than half of them. Using orgasm to help alleviate migraine pain is not as reliable as prescription medications, but it does work much faster, is cheaper, has fewer side effects, and is more fun.

Depression Treatment

Orgasms can also have an antidepressant effect. Orgasms cause intense increased activity in the deep limbic parts of the brain, which settle down after sex. Antidepressants tend to calm activity in the limbic parts of the brain as well. People who engaged in regular sexual activity experienced less depression; orgasm frequency may be one reason why. When a man has an orgasm, an area in the limbic system, called the meso-diencephalic junction, is activated. Cells in the region are known to produce some of the pleasure chemicals discussed earlier. At the same time, researchers have shown that the amygdala, a fear center in the brain, becomes less active in men's brains during sex. The region is also involved in vigilance, so animals and people may need to shut down that part of the brain to avoid getting distracted during sex. Calming the fear center might also help with a man's sense of commitment. Prostaglandins, fatty acids found in semen, are absorbed by the vagina and may have a role in modulating female hormones and moods. In one study it was reported that women who perform oral sex on their mates are less likely to suffer from preeclampsia, a condition that causes a dangerous spike in women's blood pressure during pregnancy. Plus, sperm carries TGF beta, a molecule that can boost the activities of her natural killer cells, which attack the rogue cells that give rise to tumors.

A man's orgasm can even be beneficial to women, according to research that indicates that semen can reduce depression in women. Gordon Gallup, a psychologist at the State University of New York, headed a study that found women whose male partners did not use condoms were less subject to depression than those whose partners did. One theory put forth was that prostaglandin, a hormone found in semen, may be absorbed in the female genital tract, thus modulating female hormones.

Other research has indicated that high sexual activity is associated with lower risk and incidence of depression and suicide. A Canadian study that examined the correlation between sexuality and mental health found that celibacy was correlated with high scores on depression and suicidality indexes.

Look Younger

Regular orgasms can even help you look younger. According to research done by David Weeks, a clinical neuropsychologist at the Royal Edinburgh Hospital, making love three times a week in a stress-free relationship can make you look ten years younger. Dr. Weeks studied more than 3,500 men and women between the

THE BRAIN IN LOVE · 19

ages of 18 and 102. He concluded that genetics were only 25 percent responsible for how young we look—the rest is due to behavior. In his study, a panel of judges viewed the participants through a one-way mirror and then guessed the age of each subject. A group of men and women were labeled "superyoung" whose ages were underestimated by seven to twelve years. Among these "superyoung" people, one of the strongest correlates of youthful appearance was an active sex life. They reported having sex at least three times per week, in comparison with the control group's average of twice a week. The "superyoung" were also found to be more comfortable and confident regarding their sexual identity. Dr. Weeks, whose findings are published in *Superyoung: The Proven Way to Stay Young Forever* (Hodder and Stoughton), says this is partly because sexual activity in women helps to trigger the production of a human growth hormone that helps them maintain their youthful looks. Sexual activity also pumps oxygen around the body, boosting the circulation and the flow of nutrients to the skin. Moreover, being in a sexual relationship can in itself be a good incentive to look after your appearance and stay in shape.

Improved Sense of Smell

After sex, production of the hormone prolactin surges. This in turn causes stem cells in the brain to develop new neurons in the brain's olfactory bulb, its smell center, improving one's sense of smell.

Weight Loss, Overall Fitness

One of the most compelling benefits of sex comes from studies of aerobic fitness. It has been estimated that the act of intercourse burns about two hundred calories, the equivalent of running vigorously for thirty minutes. Most couples average about twenty-four minutes for lovemaking. During orgasm, both heart rate and blood pressure typically double, all under the influence of oxytocin. Muscular contractions during intercourse work the pelvis, thighs, buttocks, arms, neck, and thorax. *Men's Health Magazine* has gone so far as to call the bed the single greatest piece of exercise equipment ever invented.

The Key to Health and Longevity

Regular sexual contact, especially with a committed partner, helps to keep your body and brain healthy. Do not use excuses such as you are too tired or too busy for physical affection. Also, try to avoid spending too much time at work at the exclusion of social endeavors. A lack of relationships sets up humans to be depressed or to seek pleasure through solitary sexual activities, such as using the Internet, drugs or alcohol, gambling, or other addictions, which are not good for the brain. Men and women need touching, eye contact, and sexual connection to stay healthy. When you feel loved, nurtured, cared for, supported, and intimate, you are much more likely to be happier and healthier. You have a much lower risk of getting sick and, if you do, a much greater chance of surviving.

Happiness

There is happy news for people who have more activity in the bedroom than in their bank accounts. After evaluating the levels of sexual activity and happiness in sixteen thousand people, Dartmouth College economist David Blachflower and University of Warwick in England professor Andrew Oswald found that sex so positively influenced happiness that they estimated increasing intercourse from once a month to once a week is equivalent to the happiness generated by getting an additional $50,000 in income for the average American. In addition, they also reported, that despite what most people think, people who make more money do not necessarily have more sex. There was no difference, in their study, between sexual frequency and income levels. The happiest people in the study were married people who had, on average, 30 percent more sex than single folks. The economists estimated that a lasting marriage equated to the happiness generated by an extra $100,000 annually, while divorce depleted an estimated $66,000 annually worth of happiness. Taking care of your marriage can save you lots of money.

Table 1: Summary of Some of the Health Benefits of Regular Sexual Contact

For women, in research studies, regular sex with a partner has been associated with:

- more regular menstrual cycles
- more fertile menstrual cycles
- lighter periods
- better moods
- better memories
- pain relief
- better bladder control
- fewer colds and flu
- reduced stress
- staying in shape
- increase in the youth-promoting hormone DHEA
- increased testosterone and estrogen
- better weight control—sex burns about 200 calories per half hour, yoga 114, dancing (rock) 129, walking (3 mph) 153, weight training 153.

For men, regular sex with a partner has been associated with:

- increased heart rate variability (a sign of heart health and a calmer mind)
- improved heart cardiovascular function (three times a week decreased risk of heart attack or stroke by half)

- higher testosterone levels (stronger bones and muscles)
- improved prostate function
- improved sleep.

Lesson #1: Remember Methuselah's secret—frequent sexual activity is good for your health.

NO FORETHOUGHT
EQUALS NO FOREPLAY
———
Understanding and Optimizing the
Brain Systems of Sex

"The great sins of the world take place in the brain: but it is in the brain that everything takes place. . . . It is in the brain that the poppy is red, that the apple is odorous, that the skylark sings."

—OSCAR WILDE

Are you:

Impulsive or thoughtful?
Rigid or flexible?
Anxious or confident?
Negative or hopeful?
Short-tempered or patient?
Able to admit problems or in denial?
Coordinated or prone to bumping into walls?
Attached or afraid?
Faithful or a wanderer?

The inner workings of the brain influence all we do sexually. Once thought of as a black box too complex to understand, the brain is now being studied and understood like never before. We know that within the brain there are systems that work together to produce our personality, cares, dreams, aspirations, and sexual competencies. In my work as a neuroscientist, I have found it useful to think about five different brain systems that relate to human behavior: prefrontal cortex, anterior cingulate gyrus, deep limbic system, basal ganglia, and temporal lobes. Each of these systems interacts with other areas of the brain to produce the effective or not-so-effective human behavior.

The brain is divided into four main lobes or regions: frontal (forethought and judgment), temporal (memory and mood stability), parietal (sensory processing and direction sense), and occipital lobes (visual processing). There are also important structures deep in the brain, such as the anterior cingulate gyrus (gear shifter), basal ganglia (anxiety and pleasure center), and deep limbic system (emotional center). A useful generalization about how the brain functions is that the back half—the parietal, occipital, and back part of the temporal lobes—takes in and perceives the world, and determines what is sexy to us. The front half of the brain integrates this information, analyzes it, decides what to do, then plans and executes the decision, such as "Shall I go on a date with him if he asks?"

This chapter will give a detailed look at the five brain systems of behavior, including a discussion of the functions, problems, and treatments of each system, especially as they relate to sexual relationships. In addition, I'll include a section on common things partners say when these systems are out of whack. This section will help readers more clearly identify individual vulnerabilities and problems. For each of the brain systems discussed, I will add an "at a glance" summary chart highlighting the major functions, problems, and treatments. See my book *Change Your Brain, Change Your Life* for more information on each system. Appendix A is a self-test to help readers evaluate these systems for themselves. Obviously, any medication recommendations need to be discussed with your doctor.

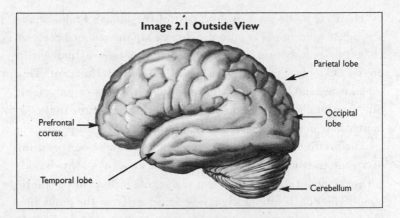

Image 2.1 Outside View

Parietal lobe

Occipital lobe

Prefrontal cortex

Temporal lobe

Cerebellum

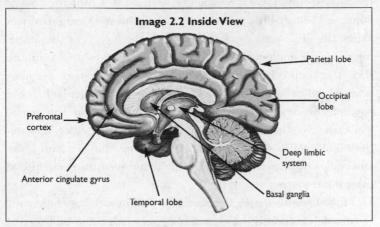

Image 2.2 Inside View

Parietal lobe

Occipital lobe

Prefrontal cortex

Deep limbic system

Anterior cingulate gyrus

Basal ganglia

Temporal lobe

Prefrontal Cortex (PFC): No Forethought Equals No Foreplay

The frontal lobes (the front half of the brain) are divided into three areas: the motor cortex, which controls the body's motor movements, such as walking, chewing, and moving your fingers and toes; the premotor area, which is involved in planning motor movements; and the prefrontal cortex (PFC), the front third of the brain, which is involved with executive functions such as planning, forethought, judgment, organizing, impulse control, and learning from past mistakes.

The PFC is the most evolved part of the human brain, representing 30 percent of the cortex, compared to the chimpanzee, our closest primate cousin, whose PFC occupies only 11 percent; a dog's PFC, only 7 percent; or a cat's PFC, only 3.5 percent. This explains my cat Annabelle, who has no forethought or judgment. She lives totally in the moment and will drink out of the toilet, no matter how many times she has been told "No."

The prefrontal cortex houses our ability to guide our behavior over time to reach our goals. When the PFC works as it should, we are thoughtful, empathic, expressive, organized, and goal oriented. The PFC is often called the executive part of the brain, like the boss at work. When it is low in activity, it is as if the boss is gone, so there is little to no supervision and nothing gets done. When the PFC works too hard, it is as if the boss is micromanaging everyone, and people are left with anxiety and worry. I call the PFC the Jiminy Cricket part of the brain. It houses our conscience and our ability to stay on track toward our goals. It is the part of the brain that, as Jiminy Cricket says in the movie *Pinocchio*, "is the still, small voice that helps you decide between right and wrong." In the dating, relationship, and sex world, the PFC helps us be patient, thoughtful, goal driven, and empathetic toward our partner.

Problems with the PFC result in a "Jiminy Cricket Deficiency Syndrome": a diminished conscience, poor judgment, impulsivity, desire to seek excitement, short attention span, disorganization, trouble learning from experience, poor time management, and lack of empathy. It has been associated with attention-deficit/hyperactivity disorder (ADHD), antisocial personality disorder, sexual addictions, brain injuries, and some forms of dementia. Low activity in this part of the brain is often due to a deficiency in the neurotransmitter dopamine; increasing it through supplements or medications is often helpful.

Healthy activity in the PFC is associated with conscientiousness; abnormal PFC activity is associated with inconsistency and troubled decisions. In reviewing 194 studies, Drs. T. Boggs and B. W. Roberts from the Department of Psychology at the University of Illinois at Urbana-Champaign found that increased death rates were associated with poor PFC activity due to impulsive behaviors (a lack of conscientiousness)—tobacco use, diet and activity patterns, excessive alcohol use, drug use, violence, risky sexual behavior, risky driving, and suicide. You need a good PFC to live long and be happy!

PFC in Relationships

When the PFC functions properly, people are able to engage in goal-directed behavior and effectively supervise their words and deeds. They are able to think before they say things and they tend to say things that effect their goals in a positive way. They also tend to think before they do things and learn from mistakes. In addition, they are able to focus and attend to conversations, follow through on commitments and chores, and organize their actions and spaces. They are also able to be settled and sit still. They are able to express what they feel. And they tend not to like conflict, tension, and turmoil.

When the PFC is underactive, people tend to be impulsive in what they say or do, often causing serious problems in relationships (such as saying hurtful things without forethought). They tend to live in the moment and have trouble delaying gratification (I want it now). They also have trouble listening in relationships and tend to be easily distracted. There is often difficulty expressing thoughts and feelings; partners often complain of a lack of talking in the relationship. They tend to be restless and fidgety. In addition, they tend to be sensitive to noise and touch. Organization of time and space is difficult and they have trouble staying on task and finishing projects, commitments, and chores. There is also a tendency to be late. In addition, many people with PFC problems have an unconscious tendency to be conflict seeking or to look for problems when none exist. I call this tendency "the game of let's have a problem." They also tend to seek stimulation or do high-risk activities that upset or frighten their partners (such as driving too fast, skydiving, getting in the middle of a fight between strangers). Also, many people with PFC problems cannot filter out noise, smells, or light and tend to be overly sensitive to their environments, thus easily distracted.

POSITIVE PFC RELATIONAL STATEMENTS

You're important to me. Let's do something tonight.

I love you. I'm glad we're together.

I love to listen to you.

I'll be on time for our date.

Let's get these chores done so that we'll have more time together.

I don't want to fight. Let's take a break and come back in ten minutes and work this out.

I made that mistake before. I'm not making it again.

I planned a wonderful evening for us.

How do you like me to touch you? I want to please you.

NEGATIVE PFC RELATIONAL STATEMENTS

I'm just a half hour late. Why are you so uptight about it?

If you want the checkbook balanced, do it yourself.

I'll do it later.

I find it hard to listen to you.

Go ahead and talk to me. I can listen to you while I'm watching TV and reading this book.

I can't express myself.

My mind goes blank when I try to express my feelings.

I didn't mean to have the affair (or, overspend, embarrass you at the party, say the hurtful comments, etc.).

I just can't sit still.

The noise bothers me.

I get so distracted (while listening, during sex, when playing a game, etc.).

I need the answer now.

I want it now.

I'm so mad at myself. I've made that mistake too many times.

STATEMENTS FROM PARTNERS OF PEOPLE WITH PFC PROBLEMS

He's impulsive.

She blurts out and interrupts.

He doesn't pay attention to me.

She won't let me finish a comment. She says she has to say a thought that comes into her head or she'll forget it.

He has to have the fan on at night to sleep. It drives me crazy.

She often seems to start a problem for no particular reason.

He loves to challenge everything I say.

She gets so distracted during sex.

He teases the animals and it makes me furious.

She can't sit still.

He puts things off and tends not to finish things.

She's always late, rushing around at the last minute.

PREFRONTAL CORTEX (PFC) SUMMARY
(the boss in your head, supervising your life)

PFC Functions (supervision)	Low PFC Problems (lack of supervision)
Focus	Short attention span
Forethought	Lack of clear goals or forward thinking
Impulse control	Impulsivity
Organization	Disorganization
Planning, goal setting	Procrastination
Judgment	Poor judgment
Empathy	Lack of empathy
Emotional control	Failure to give close attention to detail
Insight	Lack of insight
Learning from mistakes	Trouble learning from mistakes
	Tendency to lose things
	Easy distraction

DIAGNOSTIC PROBLEMS ASSOCIATED WITH LOW PFC ACTIVITY

ADHD	Some types of depression
Brain trauma	Dementia, associated with bad judgment
Antisocial personality	Conduct disorders

DIAGNOSTIC PROBLEMS ASSOCIATED WITH EXCESSIVE PFC ACTIVITY

Overfocused, rigid, and inflexible

Also, see problems of the anterior cingulate gyrus (page 34)

WAYS TO BALANCE LOW PFC

Organizational help, coaching	Intense aerobic exercise (boosts blood flow)
Goal-setting/planning exercises	Neurofeedback to boost PFC activity
Relationship counseling	Stimulating or exciting activities
Higher protein diet	Developing a deep sense of personal meaning

Stimulating supplements, to boost dopamine to the brain, such as L-tyrosine or SAMe

Stimulating medications (if appropriate), such as Adderall, Dexedrine, Ritalin, Wellbutrin, Stratterra, or Provigil

WAYS TO BALANCE HIGH PFC

See prescriptions of the anterior cingulate gyrus (page 35)

Anterior Cingulate Gyrus (ACG)—My Way or the Highway

The ACG helps you feel settled, relaxed, and flexible. It runs lengthwise through the deep parts of the frontal lobes and is the brain's major switching station. I think of it as the brain's gear shifter, greasing human behavior, and allowing us to be flexible, adaptable, and to change as change is needed. This part of the brain is involved in helping shift attention from thing to thing, moving from idea to idea, and seeing the options in life. The term that best relates to the ACG is *cognitive flexibility*. Cooperation is also influenced by this part of the brain. When the ACG works in an effective manner, it is easy to shift into cooperative modes of behavior.

When there is too much activity in the ACG, usually due to lower serotonin levels, people become unable to shift their attention and become rigid, cognitively inflexible, overfocused, anxious, and oppositional. When it works too hard, people have difficulty shifting attention and get stuck in ineffective behavior patterns, where they may be uncooperative or difficult, trapped in their own mindset. When the ACG works too hard, people plan too much, worry too much about the future, and become too serious or obsessed. Difficulties in the ACG can cause a person to constantly expect negative events and feel very unsafe in the world. When the ACG is overactive, people have a tendency to get stuck or locked into negative thoughts or behaviors. They may become obsessive worriers or hold onto hurts or grudges from the past. They may also get stuck on negative behaviors, or develop compulsions such as hand-washing or excessively checking locks. One patient who had ACG problems described the phenomenon to me saying it was "like being on a rat's exercise wheel, where the thoughts just go over and over and over." Another patient told me, "It's like having a Reset button in your head that is always on. Even though I don't want to have the thought anymore, it just keeps coming back."

The clinical problems associated with excessive ACG activity include obsessive compulsive disorder, eating disorders, and addictive disorders. All of these disorders are associated with problems shifting attention. Worrying, holding onto hurts from the past, cognitive inflexibility, automatically saying no, and being rigid, are symptoms of too much activity in the ACG.

Increasing serotonin through supplements or medications is often helpful.

When the ACG is underactive, people have little motivation and get-up-and-go. They shift gears too easily and can be easily distracted and apathetic. Neurosurgeons at the University of California at Irvine School of Medicine described a set of symptoms associated with damage to this part of the brain, from a stroke, tumor, or brain injury. They called the syndrome akinetic mutism, where patients tend to have little physical movement (akinetic) and produce little speech (mutism). They also noted indifference and lower levels of anxiety and worry in patients.

ACG in Relationships

When the ACG functions properly, people are able to shift their attention easily. They tend to be flexible and adaptable. They are likely to see options in tough situations. They are usually able to forgive the mistakes of others and tend not to hold onto hurts from the past. They encourage others to help but do not rigidly control situations. They tend to have a positive outlook and see a hopeful future. Basically, they are able to roll with the ups and downs of relationships.

When the ACG is <u>overac</u>tive, people have a tendency to get locked into thoughts, stuck on thoughts, and get the same thoughts in their head over and over. They tend to hold grudges, hold onto hurts from the past, and to be unforgiving of perceived wrongdoings. They tend to be inflexible, rigid, and unbending. They often want things done a certain way (their way) and they may get very upset when things do not go their way. They have difficulty dealing with a change. They tend to be argumentative and oppositional. When the ACG is underactive, people can act in apathetic, indifferent, or uncaring ways. Interestingly, one of the treatments for excessive activity in the ACG with disorders like obsessive-compulsive disorder is selective serotonin reuptake inhibitor (SSRI) medications such as Prozac or Lexapro. One of the side effects of these medications is apathy and lack of motivation when it suppresses the ACG too much.

Positive ACG Relational Statements

It's okay.

I can roll with this situation.

How would you like to do this?

Let's collaborate.

Let's cooperate.

What would you like to do?

That was in the past.

Negative ACG Relational Statements

You hurt me years ago.

I won't forgive you.

It'll never be the same.

I'm always worried.

I get stuck on these bad thoughts.

Do it my way.

I can't change.

It's your fault.

I don't agree with you.

No. No. No.

I won't do it.

I don't want to do it.

I have a lot of complaints about you.

I've never hated anyone more than you.

This will never change.

Statements from Partners of People with ACG Problems

Nothing ever gets forgiven or let go.

She brings up issues from years and years ago.

Everything has to be the way he/she wants it.

He can't say he's sorry.

She holds onto grudges forever.

He never throws anything away.

She's rigid.

If things aren't perfect, he thinks they are no good at all.

I don't help her because I have to do it exactly her way or she
goes ballistic.

He argues with everything I say.

She tends to be oppositional.

He doesn't like to try new things.

Anterior Cingulate Gyrus (ACG) Summary
(the brain's gear shifter)

ACG Functions	Excessive ACG Activity Problems
Cognitive flexibility	Gets stuck on negative thoughts or behaviors
Transition from idea to idea	Worries
Cooperation	Holds grudges
Ability to see options	Has obsessions/compulsions
Goes with the flow	Is inflexible, may appear selfish
	Is oppositional/argumentative
	Gets upset when things do not go his way
	Gets upset when things are out of place
	Has an intense dislike for change
	Tends to say no without thinking

Low ACG Activity Problems

Apathy or indifference

Poor motivation

Little speech production

Decreased physical movement

Diagnostic Problems Associated with Excessive ACG Activity

Obsessive compulsive disorder	Addictions

Eating disorders

Chronic pain

Oppositional defiant
disorder

Tourette's syndrome

Premenstrual tension syndrome
(some types)

Posttraumatic stress disorder

Difficult temperaments

DIAGNOSTIC PROBLEMS ASSOCIATED WITH LOW ACG ACTIVITY

Lowered motivation, little spontaneous movement or speech
See problems of low PFC activity (see page 29–30)

WAYS TO CALM EXCESSIVE ACG ACTIVITY

Neurofeedback to calm ACG activity (uses measuring instru-
ments to give people information or feedback on the activity
in this part of the brain so they can learn to calm it down)
Intense aerobic exercise
Relationship counseling, anger management
Lower protein/complex carbohydrate diet

ACG supplements, to boost serotonin to the brain, such 5-HTP,
St. John's wort, or Inositol

ACG medications (if appropriate), SSRIs (Paxil, Zoloft, Celexa, Prozac, Luvox, Lexapro), Effexor, atypical antipsychotics
in refractory cases, such as Risperdal, Zyprexa, or Geodon

WAYS TO BOOST LOW ACG ACTIVITY

See prescriptions of the PFC (page 30)

Use Practical Neuroscience to Enhance Love

A friend of mine recently approached me at a seminar and told me
this story. "Daniel, I am so grateful for everything you have taught
me, especially about the anterior cingulate gyrus. I am married to a
woman who has the anterior cingulate from hell. No matter what I
say, she says the opposite. It has been so frustrating. For years I just

thought that she didn't love me. Now I know it has to do with how her brain works. If I asked her to go to the store with me, she would always say, 'I am too busy to go. It is so insensitive of you to ask me; don't you see everything I am doing.' Rebuffed, I stopped asking her to do things. Since I have listened to you talk about the anterior cingulate gyrus, I realize that her brain gets stuck and I need to ask the opposite of what I want. For example, if I want her to go to the store with me, I'll say, 'I am going to the store. You probably do not want to go with me.' Incensed, she says, 'Of course I want to go with you. What would ever give you that idea?' We are doing much better now. But I still have one problem. It's the sex thing. It doesn't sound right to say, 'I am going to have sex. You probably do not want to come with me?' Do you have any ideas on how to solve that problem?"

In response to my friend, I smiled and said, "I am very pleased you have learned practical neuroscience to improve your relationship with you wife. I have several ideas on how to get more sex with people like your wife who have anterior cingulate gyrus problems. First, take her out for a pasta dinner. Simple carbohydrates boost serotonin levels in the brain and help people feel more relaxed and more flexible. Next, take her for a long walk. Exercise boosts blood flow to the brain and also increases serotonin levels as well. Next, when you get home, give her a small piece of Godiva chocolate, which increases a chemical called phenylethylamine (PEA), which boosts the brain's alerting system. Then rub her shoulders and never ask for anything directly. Odds are from day four to day twenty of her menstrual cycle you are likely to get lucky." Several weeks later I opened my e-mail and found a note from my friend with a string of hundreds of "thank you" phrases.

Understanding the different brain systems, such as the anterior cingulate gyrus, allows you to develop specific strategies to enhance your sex life. When someone has low PFC activity, for example, they will need to be excited or turned on in order to be more interested in sex. Taking them to a meditation session or a professional lecture is not likely to turn them on. They are more likely to need a scary movie or a ride on a motorcycle in order to get excited.

Deep Limbic System (DLS)—Passion Fires Burning

The DLS lies near the center of the brain. About the size of a walnut, this part of the brain is involved in setting a person's emotional tone. When the DLS is less active, there is generally a positive, more hopeful state of mind. When it is heated up, or overactive, negativity can take over. Due to this emotional shading, the DLS provides the filter through which you interpret the events of the day; it tags or colors events, depending on the emotional state of mind. The DLS, including structures called the hippocampus and amygdala, have also been reported to store highly charged emotional memories, both positive and negative. The total experience of our emotional memories is responsible, in part, for our emotional tone. Stable, positive experiences enhance how we feel. Trauma and negative experiences set our brain in a negative way.

The DLS controls the sleep and appetite cycles of the body and is intimately involved with bonding and social connectedness. This capacity to bond plays a significant role in the tone and quality of our moods. The DLS also directly processes the sense of smell. Because your sense of smell goes directly to the deep limbic system, it is easy to see why smells can have such a powerful impact on our feeling states.

Too much activity in the DLS is associated with depression; negativity; and low motivation, libido, and energy. Because sufferers feel hopeless about the outcome, they have little willpower to follow through with tasks. Since the sleep and appetite centers are in the DLS, disruption can lead to changes in habits, which may mean an inclination to too much or too little of either. For example, in typical depressive episodes, people have been known to lose their appetites and to have trouble sleeping despite being chronically tired. High activity in the DLS may be due to deficiencies in the neurotransmitters norepinephrine, dopamine, or serotonin;

increasing these chemicals through supplements or medications may be helpful. Low activity in the DLS has been associated with lowered motivation, decreased reactiveness, and misreading incoming information.

DLS in Relationships

When the limbic system functions properly, people tend to be more positive and more able to connect with other people. They tend to filter information in an accurate light and they are more likely to give others the benefit of the doubt. They are able to be playful, sexy, and sexual, and they tend to maintain and have easy access to positive emotional memories. They tend to draw people toward them with their positive attitude. When the limbic system is overactive, there is a tendency toward depression, negativity, and distance from others. They tend to focus on the most negative aspects of others, filter information through dark glasses, and see the glass as half empty. They tend not to be playful. They do not feel sexy and they tend to shy away from sexual activity due to a lack of interest. Most of their memories are negative and it is hard to access positive emotional memories or feelings. They tend to push people away with their negativity.

POSITIVE DLS RELATIONAL STATEMENTS

We have great memories.
Let's have friends over.
I accept your apology. I know you were just having a bad day.
Let's have fun.
I feel sexy. Let's make love.

NEGATIVE DLS RELATIONAL STATEMENTS

Don't look at me in a negative way.
All I can remember is the bad times.
I'm too tired.

Leave me alone. I'm not interested in sex.
You go to bed. I can't sleep.
I don't feel like being around other people.
I don't want to hear you're sorry. You meant to hurt me.
I'm not interested in doing anything.

STATEMENTS FROM PARTNERS OF PEOPLE WITH DLS PROBLEMS

She's negative.
He's often depressed.
She looks on the negative side of things.
He doesn't want to be around other people.
She tends to take things the wrong way.
He's not interested in sex.
She can't sleep.
There's little playfulness in our relationship.

DEEP LIMBIC SYSTEM (DLS) SUMMARY
(the mood and bonding center)

DLS Functions	Excessive DLS Activity Problems
Mood control	Depression, sadness
Memories	Focus on the negative, irritability
Degree of motivation	Low motivation and energy
Emotional tone	Negativity, blame, guilt
Appetite/sleep cycles	Poor sleep and appetite
Bonding	Social disconnections/isolation
Sense of smell	Low self-esteem
Libido	Low libido
Flight-or-fight response	Hopelessness
	Decreased interest in things that are usually fun
	Feelings of worthlessness or helplessness
	Feeling dissatisfied or bored
	Crying spells

Low DLS Activity Problems
Decreased reactiveness
Misreading incoming information

DIAGNOSTIC PROBLEMS ASSOCIATED WITH HIGH DLS ACTIVITY
Depression   Cyclic mood disorders
Pain syndromes

DIAGNOSTIC PROBLEMS ASSOCIATED WITH LOW DLS ACTIVITY
None

WAYS TO CALM HIGH DLS ACTIVITY

Biofeedback, increase left PFC activity (helps calm the DLS through its connections)
Intense aerobic exercise
Relationship counseling
Therapy to correct and eliminate ANTs (automatic negative thoughts)
Balanced diet, such as described by Barry Sears in *The Zone Diet*

DLS supplements, such as DL phenylalanine, SAMe, L-tyrosine

DLS medications (if appropriate), antidepressants such as Wellbutrin (bupropion), Effexor (venlafaxine), Norpramin (desipramine), Tofranil (imipramine), SSRIs (if ACG also present), anticonvulsants/Lithium to help with cyclic mood changes

Basal Ganglia (BG)—You Make Me Nervous

The basal ganglia are a set of large structures toward the center of the brain that surround the deep limbic system. The BG are involved with integrating feelings, thoughts, and movement, which is why you jump when you get excited or freeze when you are scared. In our clinic we have noticed that the basal ganglia are involved with setting the body's idle or anxiety level. When the BG work too hard, people tend to struggle with anxiety and physical stress symptoms, such as headaches, intestinal problems, and muscle tension. High BG activity is also associated with conflict-avoidant behavior. Anything that reminds them of a worry (such as confronting an employee who is not doing a good job) produces anxiety; high BG people tend to avoid conflict, because it makes them feel uncomfortable. People with high BG activity also have trouble relaxing and tend to overwork. When the BG are low in activity, people tend to have problems with motivation and attention.

In addition, the BG are involved with feelings of pleasure and ecstasy. Cocaine works in this part of the brain. High activity in this part of the brain is often due to a deficiency in the neurotransmitter GABA; increasing it through supplements or medications is often helpful.

BG in Relationships

When the basal ganglia system functions properly, people tend to be calm and relaxed. They tend to predict the best and, in general, see a positive future. Their bodies tend to feel good, and they are physically free to express their sexuality. They are not plagued by multiple physical complaints. They tend to be relaxed enough to be playful, sexy, and sexual. They are able to deal with conflict in an effective way. When the basal ganglia is overactive, there is a tendency toward anxiety, panic, fear, and tension. They tend to focus on negative future events and what can go wrong in a situation. They filter information through fear and they are less likely to give others the benefit of the doubt. They tend to have headaches, backaches, and a variety of physical complaints. They have lowered sexual interest because their physical bodies tend to be wrapped in tension. They often do not have the physical or emotional energy to feel sexy or sexual and they tend to shy away from sexual activity. Most of their memories are filled with anxiety or fear. They tend to wear out people by the constant fear they project.

POSITIVE BG RELATIONAL STATEMENTS

I know things will work out.

I can speak out when I have a problem. I don't let problems fester.

I usually feel physically relaxed.

I'm usually calm in new situations.

NEGATIVE BG RELATIONAL STATEMENTS

I know this isn't going to work out.

I'm too tense.

I'm scared.

I'm too afraid to bring up problems. I tend to avoid them.

I can't breathe. I feel really anxious in this situation.

I can't make love—I have a headache (chest pain, backache, muscle aches, etc.).

You're going to do something to hurt me (predict fear).

STATEMENTS FROM PARTNERS OF PEOPLE WITH BG PROBLEMS

She's anxious.

He's nervous.

She's uptight.

He cares too much about what others think.

He predicts the worst possible outcomes to situations.

She complains for feeling bad a lot (headaches, stomachaches).

He won't deal with conflict.

She won't deal with problems head on.

BASAL GANGLIA (BG) SUMMARY
(the anxiety center)

BG Functions	Excessive BG Activity Problems
Integrates feeling, thoughts and movement	Anxiety/panic
	Hypervigilance
	Muscle tension
Sets body's idle	Conflict avoidance

Smooths movement
Modulates motivation
Mediates pleasure

Prediction of the worst
Excessive fear of being judged by
others
Tendency to freeze in anxiety
situations
Shyness or timidity
Tendency to bite fingernails or
pick skin
Excessive motivation
(can't stop working)

Low BG Activity Problems
Low motivation
Attentional problems
Excitement seeking
Tremor/movement problems

DIAGNOSTIC PROBLEMS ASSOCIATED WITH EXCESSIVE BG ACTIVITY

Anxiety disorders
Physical stress symptoms,
such as headaches,
stomachaches

Workaholism
Insecurity

DIAGNOSTIC PROBLEMS ASSOCIATED WITH LOW BG ACTIVITY

Movement disorders
Low motivation

Attention deficit disorder (ADD)

WAYS TO CALM HIGH BG ACTIVITY

Body biofeedback

Hypnosis, meditation
Relaxing music
Limited caffeine/alcohol

Cognitive therapy to kill the bad
thoughts
Relaxation training
Assertiveness training

High BG supplements, such as GABA or valerian root

High BG medications (if appropriate), antianxiety meds such as benzodiazepines (low dose, short time), Buspar, antidepressant meds, anticonvulsants, blood pressure meds such as propranolol

WAYS TO STIMULATE LOW BG ACTIVITY

Intense aerobic exercise   Stimulating, exciting behaviors
Stimulating music

Low BG supplements, to boost dopamine, such as L-tyrosine

Low BG medications (if appropriate), stimulants such as Adderall or Concerta

Temporal Lobes (TLs)—Memories and Moods

The temporal lobes, underneath your temples and behind your eyes, are involved with language (hearing and reading), reading social cues, short-term memory, getting memories into long-term storage, processing music, tone of voice, and mood stability. They also help with recognizing objects by sight and naming them. Called the "What Pathway" in the brain, it is involved with recognition and naming objects and faces. In addition, the temporal lobes, especially on the right side, have been implicated in spiritual experience and insight. Experiments that stimulate the right temporal lobe have demonstrated increased religious or spiritual experiences, such as feeling God's presence. Orgasms have been found to activate this part of the brain as well.

The hippocampus, situated on the inside aspect of the temporal lobes, encodes new information and stores it for up to several weeks. When these areas are damaged, you can neither store new experiences nor retrieve experiences learned within the past several weeks. The hippocampus is one of the first areas damaged by Alzheimer's disease.

In front of the hippocampus on the inside of the temporal lobe is an almond-shaped structure called the amygdala. The amygdala coordinates your emotional responses. Strong emotions can improve the encoding process of hippocampal neurons and make it easier to retrieve the experience. This is useful because it allows you to more easily remember events that were "emotionally stimulating," such as being mugged, having good sex, or recalling a fascinating fact you recently heard. I remember a taxi ride in 2005 from Manhattan to JFK airport like it was yesterday. I even remember the cab number, 4118. The cab driver had on very irritating music, talked loudly on his cell phone, paid little attention to the road, and nearly got us into two accidents. My emotional response to this terrible ride to the airport got his cab number stuck in my head. By emphasizing the memory of certain experiences over others, the amygdala allows you to respond more appropriately and quickly in the future—being able to recognize a potential mugger or dangerous cab driver ahead of time may save your life. When the amygdala functions appropriately, we tend to react to the world in a logical, thoughtful way. When it is overactive, our responses may be exaggerated for the situation. When the amygdala is underactive, we fail to read situations accurately, and our response may not match what has happened. For example, if you laugh upon hearing from your wife that her best friend died, your amygdala may not be working properly.

Trouble in the temporal lobes leads to both short- and long-term memory problems, reading difficulties, trouble finding the right words in conversation, trouble reading social cues, mood instability, and sometimes religious or moral preoccupation or perhaps a lack of spiritual sensitivity. The temporal lobes, especially on the left side, have been associated with temper problems. Abnormal (high or low) activity in this part of the brain is often due to a deficiency in the neurotransmitter GABA; balancing it through supplements or medications is often helpful.

TLs in Relationships

When the temporal lobes function properly, people tend to have emotional stability. They are able to process and understand what others say in a clear way. They can retrieve words for conversations. They tend to accurately read the emotional state of others. They have control over their tempers. They have access to accurate memories. Because of their memory, they have a sense of personal history and identity.

When the temporal lobes do not function properly, people tend to have memory struggles. They do not have clear access to their own personal history and identity. They are often emotionally labile (up and down). They tend to be temperamental and have problems with anger. They often have violent thoughts and express their frustration with aggressive talk. They often take things the wrong way and appear to be a little bit paranoid. They may have periods of spaciness or confusion and to misinterpret what is said to them.

Positive TL Relational Statements

I remember what you asked me to do.
I have a clear memory of the history of our relationship.
I feel stable and even.
I have access to the words I need to express my feelings.
I can usually tell when another person is happy, sad, mad, or bored.
I have good control over my temper.
My memory is good.

Negative TL Relational Statements

I struggle with memory.
I blow things way out of proportion.
I get angry easily. I have a bad temper.
My moods tend to be volatile.
I tend to get scary, violent thoughts in my head.

It's hard for me to read.

I often misinterpret what others say.

I tend to be too sensitive to others or feel others are talking about me.

I tend to misread the facial expressions of others.

I frequently have trouble finding the right words in a conversation.

STATEMENTS FROM PARTNERS OF PEOPLE WITH TL PROBLEMS

He can be physically or verbally very aggressive.

She's volatile.

His memory is very poor.

She misreads situations.

He's very moody and has serious mood swings.

She takes things the wrong way.

He spaces out very easily.

She doesn't seem to learn by reading something or hearing directions. You have to show her what to do.

TEMPORAL LOBES (TLs) SUMMARY
(memory and mood stability)

TL Functions	TL Problems (both under and overactive)
Understanding using language	Memory
Auditory learning	Auditory and visual-processing
Retrieval of words	Finding the right word
Emotional stability	Mood instability
Facilitating long-term memory	Anxiety for little or no reason
Left side: reading words	Headaches or abdominal pain (hard to diagnose)
Read faces	Reading facial expressions or social cues
Right side: reading social cues	Dark, evil, awful or hopeless thoughts

Verbal intonation	Aggression, toward self or others
Rhythm, music	Learning
Visual learning	Illusions (shadows, visual, or auditory distortions)
Spiritual experience	Overfocused on religious ideas

DIAGNOSTIC PROBLEMS ASSOCIATED WITH LOW TL ACTIVITY

Head injury	Dissociation
Anxiety	Temporal epilepsy
Amnesia	Serious depression with dark or suicidal thoughts
Left side: aggression, dyslexia	Right side: trouble with social cues

DIAGNOSTIC PROBLEMS ASSOCIATED WITH HIGH TL ACTIVITY

Epilepsy	Religiosity
Increased intuition or sensory perception	Same interventions as low TL activity

WAYS TO BALANCE THE TLs (LOW OR HIGH)

Biofeedback to stabilize TL function	Relationship counseling
Anger management	Music therapy
Increased protein diet	

For memory problems (options to consider)
- Physical and mental exercise
- Omega-3 fatty acids, alpha lipoic acid, Vitamin E and Vitamin C as antioxidants, phosphatidal serine, ginkgo biloba

TL supplements, such GABA or valerian to calm TLs if needed

TL medications (if appropriate), antiseizure medications for mood instability and temper problems, such as Depakote, Neurontin, Gabatril, and Lamictal; memory-enhancing medications for more serious memory problems, such as Namenda, Aricept, Exelon, or Reminyl

Lesson #2: No forethought equals no foreplay. Understand how the brain works and influences our behavior—and our intimate relationships—for a healthy and romantic sex life.

THE CHEMISTRY OF LOVE

Ingredients of Attraction, Infatuation, Commitment, and Detachment

"Don't copulate with people you don't want to fall in love with, because indeed you may do just that."

—Anthropologist HELEN FISHER

You're sitting behind the wheel of your van at a long traffic light. The only thing slower than the traffic is your perception of time's passage.

Then you notice her.

She appears at the curb, waiting to cross. No, she's not the love of your life. She's more like the heat of the moment. It's fortunate that your wife isn't there; you'd be in deep trouble as you take in the stranger's hips and breasts, and the way her waist scoops in to accentuate both. Time is enhanced; there's a pleasing buzz connecting your temples.

Your reaction is automatic, reflexive, and quite possibly the most powerful one you'll have this day. It temporarily blots out your long-range commitments—your ten-year marriage, your sweet child in second grade, your responsibility to keep eyes forward at traffic lights. You've surrendered control; you're captivated by the pleasure in the vision.

"You dog!" you may whisper under your breath, embarrassed by what you're envisioning as you sit there in your family van. But it might be more correct to say, "You dopamine fiend!"

When you spot the object of your desire, the neurotransmitter dopamine lights up areas deep within the brain, triggering feelings of pleasure, motivation, and reward. (Cocaine acts the same way.) You feel a rush, and your heartbeat quickens. Attraction, too, is a powerful drug. The brain stem also gets into the act, releasing phenylethylamine (PEA), which speeds up the flow of information between nerve cells. It's no wonder your neck and eyeballs track her every movement.

But she's not gawking back at you, and it's not just because you're driving a family bus with a paint scrape on the fender. Her brain acts very differently from yours. You're keyed in to beauty, shape, fantasy, and obsession; on some biological level that she may be unaware of, she's looking for a mate who will sire healthy children and protect and provide for her and them. And yes, maybe even buy them a family van. Her goals are programmed for the long range; yours are often shockingly short term.

The whole encounter can leave you quivering with pleasure, hoping for more.

It can also ruin your life.

Between the "walk" and "don't walk" signals of delight and disaster, your brain is sorting information, making choices, spurring actions. Pay attention—your whole life is riding on the choices you make.

Chemical Symphony

If you have ever listened to a symphony or any other beautiful musical production, you realize how each musician plays an essential part in the overall artistic performance. If any musician plays out of sync with the others, the music can be ruined. A fulfilling sexual relationship is very similar to a well-conducted symphony with respect to the synergy of many hormones and chemicals that are released at different phases of the love relationship. If a hormone or brain chemical is out of balance compared to the others (over- or underproduced), the entire sexual experience can be ruined.

After looking at the five different brain systems that relate to human behavior discussed in Lesson Two, it is time to explore the major chemicals involved with the primary phases of love: attraction, infatuation, commitment, and detachment.

1. **Attraction,** the craving for sexual gratification, is primarily driven by the male and female hormones, testosterone and estrogen, the chemical nitric oxide, and potentially a group of chemicals called pheromones.
2. **Infatuation,** intense, passionate love—characterized by extreme happiness when things are going well, bad feelings when they're not; focused attention; obsessive thinking; and craving for the new love—is controlled by a cocktail of neurotransmitters including epinephrine, norepinephrine, dopamine, serotonin, and phenylethylamine (PEA).
3. **Commitment,** the sense of connectedness, sustained joy, stability, and peace that one feels with a potential long-term partner, is created by the hormones oxytocin and vasopressin.
4. **Detachment,** losing a love through breakups or death, often leads to deficiencies in serotonin and endorphins.

This chapter will go through each phase and also give you tips on managing these phases, especially detachment, so they don't throw you out of balance.

Attraction Chemicals—"You Turn Me On" (Testosterone, Estrogen, Nitric Oxide, and Pheromones)

When I saw her for the first time, it was hard to catch my breath. She was stunning. I couldn't possibly think about anything else. I didn't want to stare, but couldn't help myself. "Stop it," I screamed inside my head. "Look away. No, I just have to look. Be polite, she'll think you are a pervert." My thoughts started to run wild. "Her curly brown hair, her sparkling green eyes, her neck, her slender body, amazing . . . Stop it, get a grip, you don't even know her." This little intense dialogue is what could be the beginning of the great love in your life, a horror story of obsession, or momentary fireworks.

What happens in the brain when we are sexually attracted to someone new? Our brain is programmed for attraction. It is one of the most powerfully rewarding reactions in this history of our species. The brain is a chemical factory looking for love. Attraction activates the erector sets that live in all of us.

Fifty percent of the brain is dedicated to vision, so how people look, how they move, how they smile, and how their eyes appear are critical to the process of attraction. When a man or woman sees an attractive person, the visual areas of the brain spark with activity, and he or she occupies a large portion of the brain. Attraction works very much like a powerful drug.

Using sophisticated imaging equipment, researchers from Emory University in Atlanta found that the amygdala, an area of the brain that controls emotions and motivation, is much more activated in men than women when viewing sexual material for thirty minutes, even though both sexes reported similar levels of interest in the images. This may be one of the reasons men are much more interested in pornography than women. It is also no mistake that women spend more time caring for their physical appearance. How they look has much more impact on a man's brain than the other way around.

Men tend to be attracted to symmetrical, fertile, healthy, younger-looking women. A man's genetic brain is looking at a woman and deciding whether or not he wants his children to carry her genes. Unconsciously we look for signs of health, such as clear skin and bright eyes. A number of scientists believe that body symmetry also plays a critical role in our view of beauty. The theory behind this notion is that asymmetrical features give clues to

underlying health problems, thus yielding more troubled offspring. In a study at the University of New Mexico, college males
found symmetrical female faces more attractive than asymmetrical faces. In addition, women who were blessed with symmetry
had a history of more sexual partners and tended to lose their vir-
ginity at an earlier age.

There is now scientific proof for something people have long
suspected—beautiful women make men stupid. Canadian researchers showed men pictures of conventionally pretty or not-so-
pretty women. The men rolled dice—and were told they could
either receive $15 the following day or $75 after waiting a few
days. The men who saw the pictures of the beautiful women were
more likely to take the $15—proving, researchers say, that men
stop thinking about long-term consequences once love chemicals
kick in. **By the way, the same test was done on women—and
attractiveness had no effect on their thinking processes.** It
seems as though beautiful women cause a man's limbic system to
fire up (emotional charge) while his prefrontal cortex (PFC) heads
south, leaving the judgment area of the brain vacant. Las Vegas
knows this principle very well. Casinos have beautiful waitresses
dressed in low-cut, short dresses serving free alcohol—both lower
PFC activity. No wonder the house has the edge.

A woman's brain is much less interested in how a man looks
than in how he thinks and acts. Women often look to a man's ability to care for her and her subsequent offspring. The trappings of
a successful man, in whatever society, are more important than
just his physical appearance. As always, beauty is in the eye of the
beholder.

The feelings of attraction, desire, arousal, and orgasm are fueled
by a complicated interplay of chemicals, hormones, and other substances that cause the familiar and intriguing sensation of being
"turned on." Testosterone and estrogen were first identified in the
1920s as playing a role in sexual attraction. Since that time, there
has been quite an evolution in the understanding of the chemistry
involved with feelings of lust, from the controversial work of Alfred

Kinsey in the 1940s, to the first published studies on human sexuality and the stages of sexual response in the 60s, to the present, when substances like Viagra (a medication that increases blood flow to the genital areas to aid with arousal) and AndroGel (a testosterone gel applied to the skin for those with low levels of testosterone) are readily used and heavily marketed.

A hormone is a chemical produced by one organ (endocrine gland) that has a specific effect on the activities of other organs in the body. The major sex hormones can be classified as androgens or estrogens. Both classes of hormones are present in males and females alike, but in vastly different amounts. Most men produce 6 to 8 mg of testosterone (an androgen) per day, compared to most women who produce 0.5 mg daily. Estrogens are also present in both sexes, but in larger amounts for women.

Androgens/Testosterone

Androgens, of which testosterone is the primary one, are sex hormones produced primarily by a male's testicles, but they are also produced in small amounts by the female's ovaries and the adrenal glands. Androgens help trigger the development of the testicles and penis in the male fetus. They jump-start the process of puberty and influence the male secondary sex characteristics, the development of facial, body, and pubic hair; deepening of the voice; and muscle development. After puberty, testosterone plays a role in the sex drive. Deficiencies of testosterone may cause a drop in sexual desire, and excessive testosterone may heighten sexual interest in both sexes. In men, too little testosterone may cause difficulty obtaining or maintaining erections. As we age, testosterone levels begin to decrease; as many as five million men suffer from abnormally low testosterone levels (a condition known as hypogonadism).

Unfortunately, most men never seek treatment either because they just think it is a normal part of aging or they are embarrassed to admit that they have a problem. Very often it is the female partner or wife who encourages her husband or boyfriend to seek help. This was the case with William, a fifty-six-year-old male who came to my office on the advice of his wife who had noticed a decrease in his sexual desire and approaches when he was typically a highly sexual partner. He noted that he still liked to cuddle with his wife and loved her dearly, but was rarely having erections upon awakening (a normal event in healthy males) and was not having as many spontaneous erections when feeling aroused. He also described a decreased interest in sex and felt less motivation and interest in other parts of his life as well. Both his blood and saliva tests revealed very low levels of testosterone; prescribing AndroGel (a testosterone gel applied to the shoulders once daily) restored his levels to normal, and his sexual interest and erectile function returned.

In women during their reproductive years, testosterone surges just before ovulation, in the middle of the menstrual cycle, lead many to a higher level of desire when they are most fertile. Some physicians believe that birth control pills may possibly be the cause of a decreased libido in many women because of how they interfere with ovulation to prevent pregnancy through the manipulation of levels of testosterone and estrogen. Low testosterone levels have also been associated with Alzheimer's disease and other memory problems, heart disease, and lowered bone density. If you have a low libido and memory problems, it is critical to have your testosterone levels checked.

Estrogens

Estrogens are the sex hormones produced primarily by a female's ovaries that stimulate the growth of a girl's sex organs, as well as her breasts and pubic hair, known as secondary sex characteristics. Estrogens also regulate the functioning of the menstrual cycle. Estrogens are important in maintaining the condition of the vaginal lining and its elasticity, and in producing vaginal lubrication. They also help preserve the texture and function of a woman's breasts. Considering when women with deficient desire are given estrogen and testosterone separately, the increase in desire is not as dramatic as when they are given the two hormones together, estrogen is thought to play a synergistic role with testosterone in increasing lustful desire.

In women and men, estrogen is also produced in the brain; though the contribution of estrogen to male sexual behavior has not been completely established yet, researchers are speculating that estrogen may also be very important for sexual appetite in men. An unusually high level, however, may reduce sexual interest, cause erectile difficulties, produce some breast enlargement, and result in the loss of body hair in some men. Unfortunately all of us are being exposed daily to a large amount of "xenestrogens," chemicals in the environment such as pesticides, which look like estrogen and bind to those receptors in the body. Also, inorganic chicken and beef may be injected with multiple hormones including estrogens to plump them up for the slaughterhouse. If you have the choice, it is better to select organic chicken and beef.

Nitric Oxide

Nitric oxide is a chemical released by the genitals when you are "turned on" that causes blood vessels to dilate and increases blood flow especially to the penis. Drugs like Viagra and Cialis work by stimulating the release of nitric oxide. Though these medications can work very well for some men, results of the studies that have been done thus far on women have not been compelling. Also, because these medications affect blood vessels, caution must be taken in giving them to individuals with blood pressure problems or heart disease.

Pheromones

Have you ever noticed how you have been attracted to the way someone of the opposite sex smells while another's scent may completely repel you? Pheromones, scented hormones secreted by sweat glands primarily in the armpits, are thought to attract the opposite sex. In 1991, a research group from Harvard University proved the existence of this "sixth sense," or human vameronasal system. How these hormones work is not clearly understood yet, but they are thought to influence how humans mate, bond, and take care of their offspring. Women in college dorms, or who spend a lot of time together, develop synchronized menstrual cycles and pheromones are thought to be responsible for this phenomenon. In primitive times, scent was one of the first methods of communication, and it is still an important part of how humans relate and who they will want sexually. According to neurologist Alan Hirsch, smell also has a tremendous impact on attraction. "If you smell good, we want you closer; if you smell bad, we want you to go away. When you think about the sexual organ, you really should be speaking of the nose." He goes on to say that we talk about love at first sight, when we really should be speaking of love at first sniff, because there's a direct connection between the olfactory bulb at the top of the nose and the septal nucleus of the brain, the erection center. Sexual arousal is also associated with engorgement of the erectile tissue in the nose. Dr. Hirsch has treated patients who have smell and taste disorders, and found that almost a quarter of people who lost their sense of smell develop sexual dysfunction. Measuring penile blood flow with what looks like a small blood pressure cuff, he found that sexual arousal in men was enhanced by the smells of lavender and pumpkin pie. Doughnuts, licorice, and cinnamon were also on the top of the list. (I will discuss this topic in more depth in Lesson Nine, which covers aphrodisiacs.)

Infatuation Chemicals—
"I Can't Get You Out Of My Head"
(Epinephrine, Norepinephrine, Dopamine, Serotonin, and Phenylethylamine)

Mother Nature formulated a very powerful concoction when she created the potion of chemicals involved with infatuation. The biology is smart because if it weren't for this forceful surge of chemicals, the inhibitory centers in the brain (primarily the amygdala), which warn us of potential danger or heartache, would prevail and people would never meet, mate, and procreate. Some have described the infatuation stage as an "altered state of consciousness" or akin to being "intoxicated" or "under the influence." People in this phase tend to sacrifice sleep, stay up late for hours talking with their lover on the phone, send abundant e-mails daily, and engage in behaviors that they wouldn't typically do, such as skydiving when they are afraid of heights or eating sushi when the thought of raw fish has always made them gag.

Romantic love and infatuation are not so much of an emotion as they are motivational drives. They are part of the brain's reward system. These feelings intensify to compel lovers to seek mating partners. The brain links these drives to all kinds of specific emotions, depending on how the relationship is going. All the while, our PFC is assembling information, putting the pieces of data into patterns, coming up with strategies, and monitoring progress toward "life's greatest prize." The chemicals that stimulate the motivation and drive system in the brain are the neurotransmitters epinephrine, norepinephrine, dopamine, serotonin, and phenylethylamine (PEA). These neurotransmitters play a role in the initial phase of attraction as well but it is really in the second phase of infatuation where their release becomes more active and predominant. Neurotransmitters are chemicals that help regulate the electrical signals between nerve cells in the brain. The brain is constantly seeking to keep itself balanced through increasing or decreasing amounts of these substances, some of which excite the body, for example when you see your new love from a distance and your heart starts to pound uncontrollably, and some of which calm the system and allow you to enjoy the moment, "smell the roses," and have the "warm, fuzzy" feelings associated with a new relationship.

Epinephrine and Norepinephrine

Epinephrine and norepinephrine, produced in the adrenal glands, spinal cord, and brain, are considered excitatory neurotransmitters because they cause that "adrenaline rush" feeling when the heart beats faster, blood pressure goes up, and the body is prepared to take action either in the face of a threat or in the presence of a positive stimuli such as a potential love partner. The feeling of zest and excitement comes from these chemicals as they help to facilitate both sexual arousal and orgasm. High levels of these chemicals are associated with anxiety, and low levels with depression. Chronic stress, low levels of the sex hormones estrogen, testosterone, and progesterone, a sedentary lifestyle, poor diet, and genetics can all lead to low levels of epinephrine and norepinephrine, creating interference with "the laws of attraction." Certain medications, such as stimulants or supplements like the amino acid tyrosine, can help to increase levels of these chemicals in those who have a deficiency. Medications and forms of therapy including hypnosis and biofeedback are also used to decrease these levels when they get too high, such as when individuals refrain from asking someone who they find attractive out on a date because they are afraid they will get overly anxious, their palms and face will get sweaty, or they will fumble over their words.

Dopamine

The most important and well-studied neurotransmitter associated with infatuation is dopamine. Produced in the central part of the brain, dopamine is associated with pleasure, motivation, and concentration. It has been shown to work in the reward centers of the brain. Proper amounts are associated with healthy motivation and sexual drive. Individuals feel "sexy" when they have enough of this chemical. A study by Dr. Helen Fisher published in 2002 helps explain the activity of dopamine in the brain when people are falling in love. She and a team of experts recruited forty subjects who had just fallen in love—twenty who stayed in love, the other half who had recently split up. She put each of these people into an MRI tube with a photo of a sweetheart and one of an acquaintance. Each subject looked at the sweetheart photo for thirty seconds, then—after a distraction task—at the acquaintance photo for another thirty seconds. They switched back and forth for twelve minutes. The result was a picture of the brain in love. There was increased activity in the right ventral-tegmental area. This is the part of the brain where dopamine cells project into other areas of the brain, including the basal ganglia, part of the brain's system for reward and motivation. The sweetheart photos, but not the acquaintance photos, caused this to happen. In addition, several parts of the prefrontal cortex that are highly wired in the dopamine pathways were used, while the amygdala in the temporal lobes, associated with fear, was temporarily put out of commission.

While high levels of dopamine are associated with attraction, low levels are associated with certain types of depression, attention-deficit/hyperactivity disorder (ADHD), and excitement-seeking or high-risk-taking behavior. Both cocaine and stimulant medications, such as methylphenidate (Ritalin), have been shown to enhance its production. Bupropion (Wellbutrin) is an antidepressant that enhances dopamine availability to the brain as well as enhances sexual function. Also, certain amino acid supplements like tyrosine can be used to increase dopamine levels and potentially sexual function as well. I have seen both women and men benefit from Wellbutrin and/or amino acid supplementation either when sexual function has been diminished from low levels of dopamine or when substances like SSRI medications have suppressed sexual interest.

Serotonin

Serotonin is known as the "feel good" neurotransmitter and is produced in the midbrain and brain stem. Satisfaction with a partner and the positive feelings after an orgasm are to a large degree controlled by serotonin. Normal serotonin levels help people have healthy moods and motivation. Serotonin is involved with mood regulation and emotional flexibility. Low serotonin levels have been associated with depression, anxiety, obsessive-compulsive disorder, impulsivity, and excessive activity in the brain's anterior cingulate gyrus (ACG). Low levels have also been associated with new love. Unfortunately, this makes lovers vulnerable to high anxiety levels and moodiness, common in the initial stages of a relationship. When serotonin levels are low and the ACG works too hard, people tend to get stuck on certain thoughts or behaviors. Remember the last time you fell in love. All you could think about was your new love, and no matter how busy you were, you could always find time for her. Your moods were up when you thought about her and then down when she didn't answer her cell phone the first time you called. You felt more reckless and your friends wondered about your judgment. Lowered levels of serotonin make you vulnerable to depression if the relationship ends prematurely.

In my clinical practice, I spend most of my time scanning the brains of people who need help. As part of my research, I scan many healthy people as well. Several years ago, one of my friends was scanned as part of our healthy-brain study. Several months later he fell madly in love. One day he dropped by my office to tell me about his new love. I could hear in his voice that he was so taken with his new woman that I did a repeat scan on him just to see his brain in love. The second scan showed significant increased activity in the anterior cingulate gyrus and basal ganglia, indicating his brain was literally obsessed with the woman; probably a measure of his serotonin levels would have revealed low levels at that time.

High serotonin levels can also be a problem and are associated with lowered motivation. Medications that enhance serotonin, such as selective serotonin reuptake inhibitors (SSRIs) like Prozac and Lexapro are notorious for decreasing sexual drive and function in part because recipients of these medications can lose interest if there is too much serotonin circulating in the brain, but also because these medications can decrease sensation in the genital areas, making it harder to achieve orgasm.

Dopamine and serotonin tend to counterbalance each other in

the brain. When dopamine levels are high, such as in new love, people tend to be motivated and driven toward dating behaviors that bring people closer together. Higher levels of dopamine cause lower levels of serotonin, which have been associated with obsessive thinking, hence the feeling of falling in love. When serotonin levels are high, people tend to have lowered motivation and an almost "I don't care" attitude.

Phenylethylamine

Phenylethylamine (PEA), an adrenalinelike substance, the chemical found in chocolate, speeds up the flow of information between nerve cells and is triggered in the process of attraction to help us pay attention to the love feelings. PEA is known as the "love molecule" because it is what initiates the flood of chemicals into the brain along with norepinephrine and dopamine to create the feelings of euphoria and infatuation when we are highly attracted to someone.

Commitment Chemicals—"I Love You" (Oxytocin and Vasopressin)

For those of us who have had the wonderful experience of falling in love and being infatuated with someone, we also know that this amazing high and trancelike bliss does not last forever. We either progress into deeper love and commitment or make the decision to break apart and detach. Neuroscientists have determined that after a period of anywhere from six months to two years, the brain downshifts its response and the production of stimulating chemicals and levels of neurotransmitters like PEA start to drop off. It is the body's innate wisdom to turn down the volume because it cannot maintain the lust-crazed state forever or people would eventually collapse with exhaustion. Several of my colleagues who do family and couples therapy note that a lot of unnecessary divorces and relationship breakups can occur during this time because people mistake the lack of intensity and euphoria as a sign that they have fallen out of love. Also because individuals may feel a withdrawal from the chemicals of infatuation, they may look for other sex partners, alcohol, or other substances to try to re-create the high. Understanding this phenomenon in advance can really help partners anticipate this phase and help them move into the next phase of trust and commitment where true love really begins.

Once you find an attractive partner, how does your brain decide if you want to keep him or her? Commitment is usually harder for men than women. Even though our goals are the same (continuity of the species, pleasure, and connection), women are more oriented to raising children. There is not one human society where men are primary caretakers for kids. Men and women are wired differently. Women have a larger limbic or emotional brain. It doesn't mean that men are not essential in childrearing or that they won't help. They just have different roles. Unless women have experienced emotional trauma, they are usually more ready to settle down and start a family. Men are often frightened by the responsibilities involved in raising children and being faithful to one woman, which is easier for men who have lower testosterone levels. An American study of over four thousand men found that husbands with high testosterone levels were 43 percent more likely to get divorced and 38 percent more likely to have extramarital affairs than men with lower levels. They were also 50 percent less likely to get married at all. Men with the least amounts of testosterone were more likely to get married and to stay married, maybe because low testosterone levels make men calmer, less aggressive, less intense, and more cooperative. The desire to commit to someone is strongly linked to two other hormones of emotional bonding, oxytocin and vasopressin.

Oxytocin

Oxytocin is released by the pituitary gland and acts on the ovaries and testes to regulate reproduction. Researchers suspect that this hormone is important for forming close social bonds. The levels of this chemical rise when couples watch romantic movies, hug, or hold hands. Prairie voles, when injected with oxytocin, pair much faster than normally. Blocking oxytocin prevents them from bonding in a normal way. This is similar in humans, because couples bond to certain characteristics in each other. This is why you are attracted to the same type of man or woman repeatedly. In general, levels of oxytocin are lower in men, except after an orgasm, where they are raised more than 500 percent. This may explain why men feel very sleepy after an orgasm. This is the same hormone released in babies during breast-feeding, which makes them sleepy as well.

Oxytocin is also related to the feelings of closeness and being "in love" when you have regular sex for several reasons. First, the skin is sensitized by oxytocin, encouraging affection and touching behavior. Then, oxytocin levels rise during subsequent touching and eventually even with the anticipation of being touched. Oxytocin increases during sexual activity, peaks at orgasm, and stays elevated for a period of time after intercourse. This may also be why men are more likely to talk and feel emotionally connected after sex. In addition, there is an amnesic effect created by oxytocin during sex and orgasm that blocks negative memories people have about each other for a period of time. The same amnesic effect occurs from the release of oxytocin during childbirth, while a mother is nursing to help her forget the labor pain, and during long, stressful nights spent with a newborn so that she can bond to her baby with positive feelings and love.

Higher oxytocin levels are also associated with an increased feeling of trust. In a landmark study by Michael Kosfeld and colleagues from Switzerland published in the journal *Nature*, intranasal oxytocin was found to increase trust. Men who inhale a nasal spray spiked with oxytocin give more money to partners in a risky investment game than do men who sniff a spray containing a placebo. This substance fosters the trust needed for friendship, love, families, economic transactions, and political networks. According to the study's authors, "Oxytocin specifically affects an individual's willingness to accept social risks arising through interpersonal interactions." The scientists studied oxytocin's influence on male college students playing an investment game. Each of fifty-eight men was paid $64 to participate in the experiment. The volunteers were paired up, and one man in each pair was randomly assigned to play the role of an investor and the other to play the role of a trustee. Each participant received twelve tokens, valued at thirty-two cents each and redeemable at the end of the experiment. The investors decided how many tokens to cede to the trustees. Both participants, sitting face to face, knew that the experimenters would quadruple that investment. The trustee then determined whether to keep the entire, enhanced pot or give some portion of the proceeds—whatever amount seemed fair—to the investor. Among the investors who had inhaled oxytocin, about half gave all their tokens to the trustees, and most of the rest contributed a majority of their tokens. In contrast, only one-fifth of the investors who had inhaled a placebo spray forked over all their tokens, and another one-third parted with a majority of their tokens. Oxytocin influenced only the investors. Trustees returned comparable amounts of money after inhaling either spray. The trustee responses were generous when the investors offered most of their tokens and were stingy when the investment was small. The influence of oxytocin was limited to social situations. The oxytocin influence is "a remarkable finding," says neuroscientist Antonio Damasio of the University of Iowa College of Medicine in Iowa City in an editorial published with the research report. Damasio had previously argued that the hormone acts somewhat as a love potion. "It adds trust to the mix, for there is no love without trust," he says.

Bonding chemicals can also enhance fertility. Increases in oxytocin have enhanced fertility in some studies in animals. In humans, increased oxytocin levels are associated with decreased stress levels and increased trust, both of which are likely to enhance conception.

Vasopressin

Other clues to male commitment come from new research on the hormone vasopressin. This chemical is involved in regulating sexual persistence, assertiveness, dominance, and territorial markings. Not surprisingly, it is found in higher levels in the male brain. Why do some men constantly live with the discomfort of a wandering eye, while others remain content with fidelity? The difference may have to do with vasopressin, which has been found in male voles (little rodents) to make the difference between stay-at-home dads and one-night-stand artists, for example. The voles with a certain brain distribution of vasopressin were monogamous, while others with a different pattern were not.

High levels of oxytocin and vasopressin may interfere with dopamine and norepinephrine pathways, which may explain why attachment grows as mad, passionate love fades. The antidote may be doing novel things together to goose the two love neurotransmitters. Elevated testosterone can suppress oxytocin and vasopressin. There is good evidence that men with higher testosterone levels tend to marry less often, be more abusive in their marriage, and divorce more regularly. The reverse can also be true. If a man holds a baby, levels of testosterone go down, perhaps in part because of oxytocin and vasopressin going up.

The trust, bonding, and persistence created by oxytocin and vasopressin are critical for a partnership to succeed. However, the release of these hormones is not enough by itself to keep two people compatible sexually and romantically. It is at this time when it is critical for partners to communicate their desires and needs to each other both in the bedroom and outside of the bedroom, to listen attentively and be mutually supportive of the bond that has formed from attraction to commitment. If you haven't seen it yet, see the movie *March of the Penguins* for one of the best demonstrations of true commitment.

Detachment Chemicals: Why It Hurts
(Serotonin and Endorphins)

When Shawna and Nick broke up, he was a mess. He couldn't stop thinking about her, hearing her voice in his head, feeling her touch on his body, and smelling her scent in his clothes. After being together for five years, everything reminded him of Shawna, from songs to pictures to movies to waking up and going to bed. She had been wrapped up in most of the thoughts of his day. A part of him was okay with her going away, in fact, even glad. They could never get on the same page in the relationship, and they had broken up several times before. Nick had the sense that she would not always be there for him and that she would go away if things got tough. Yet despite the deep ambivalence in the relationship, he was still a mess when she left. He couldn't sleep, he felt constantly anxious and unbalanced, and he even had panic attacks when the longing for Shawna overwhelmed him.

What happens in the brain when you lose someone you love? Why do we hurt, long, even obsess about the other person? When we love someone, they come to live in the emotional or limbic centers of our brains. He or she actually occupies nerve-cell pathways and physically lives in the neurons and synapses of the brain. When we lose someone, either through death, divorce, moves, or breakups, our brain starts to get confused and disoriented. Since the person lives in the neuronal connections, we expect to see her, hear her, feel her, and touch her. When we cannot hold her or talk to her as we usually do, the brain centers where she lives becomes inflamed looking for her. Overactivity in the limbic brain has been associated with depression and low serotonin levels, which is why we have trouble sleeping, feel obsessed, lose our appetites, want to isolate ourselves, and lose the joy we have about life. A deficit in endorphins, which modulate pain and pleasure pathways in the brain, also occurs, which may be responsible for the physical pain we feel during a breakup.

Getting a Loved One out of Your Head and the Fishhooks out of Your Heart

In Dean Koontz's novel *Velocity*, the killer uses fishhooks to torture Billy Wilens, a good-natured bartender who finds himself in a random storm of murder and mayhem. The killer, a twisted psychopathic performer, renders Billy unconscious and plants three fishhooks deep in his forehead. The fishhooks were extremely painful to extricate, requiring alcohol and painkillers, and they left deep scars.

When a loved one leaves us, or even when we are the ones who instigate a breakup, many of us feel like Billy Wilens. We have deep wounds, like fishhooks, that leave lasting scars. Many of us use alcohol or any sort of painkillers (such as drugs, sex, or excessive work) to medicate the pain. I have firsthand knowledge of what it takes to survive the painful loss of a love. I have left relationships and I have been left. Being left is definitely harder.

On one occasion after being left, it felt as though I had many fishhooks in my mind and my heart that were painfully pulled whenever I remembered any good thing my lost lover had done for or with me. Pictures, songs, friends, cars, names (she had a common name), cities, pillows, and restaurants all reminded me of her. I was a neurochemical mess for nearly six months. I even scanned myself in the middle of the process to see what grief looked like in my brain. It showed excessive anterior cingulate gyrus activity (not normal for me), which was partially why I felt so sad and obsessed.

From my own experience and work with patients who have lost loved ones, here are five tips to survive and eventually thrive through the loss of a love.

1. Above all, stay healthy. At first, we just want to medicate the pain. We eat or drink too much, stop exercising, wallow, and isolate ourselves. Stop that behavior immediately. Watch what you eat, exercise more, not less (exercise has been found to be as effective as the antidepressant Zoloft for depression), and spend time

with your friends. Make sure you get enough sleep. During my breakup, sleep was very hard for me. The supplement kava kava was helpful on a short-term basis.

2. Do not idealize the other person. Whenever we just focus on someone's good qualities, the pain increases; when we focus on his or her bad qualities, the pain decreases as we are glad to be rid of them. Spend time to write out the bad times and your ex's bad points. Whenever we lose someone we love, there is a tendency to exclusively remember the wonderful things about her. Idealizing people impairs the grieving process and makes us hurt more. Be balanced. You do not have to vilify them, but be honest about their bad qualities and focus on them to help soothe the pain. One helpful technique I found during the loss of a painful love was to make up a mnemonic that helped me remember her bad qualities. That way, when the fishhook memories pulled painfully at my heart, I could immediately remember, and silently repeat to myself, why I was glad she was gone. For example, if her name was Hanna, you could use the letters of her name to label the bad or irritating qualities: Hanna could stand for . . .

> **H**airy lip
> **A**rgumentative
> **N**ever able to say she was sorry
> **N**o memorable sex
> **A**mbivalent about our relationship

You may need these memory skills in the midst of grief to remember why you are happy someone is gone.

3. Cry, then hide the pictures. In the beginning of the breakup, take some time to allow yourself to feel the pain. Crying can be a wonderful release of the built-up tension in your limbic brain. But after a good cry, eliminate the constant triggers to your nervous system. Go through the house, your computer, and workplace and collect the pictures and gifts, then hide them somewhere. Hide

them initially, rather than burning them, because you never know what will happen in the future. If you get back together, you will feel terrible about having burnt them. Time will tell. A few months down the line you will make better decisions about whether or not you want to keep some of the things that represent your relationship. But, in the short term, get them out of sight.

4. Love must be tough. When you act weak, needy, or demanding during a breakup, you literally push the other person away. You are no longer attractive or appealing. You seem and act as a victim. Being well is not only the best revenge; it is the best way to be well.

5. Do "The Work." Byron Katie has written a wonderful, wise book titled *Loving What Is*. In the book, written with her husband, Stephen Mitchell, she discusses the power of asking yourself four questions and doing a turnaround. During my breakup it was the single most helpful technique to get my usually happy, healthy self back. I learned that I suffer whenever I do not accept what is. When I fight against reality, I am out of my mind. Katie teaches you to understand the thoughts that cause the suffering, such as "I miss her," and ask four questions and a turnaround.

- Question #1: Is it true? You bet. I miss her terribly.
- Question #2: Is it *absolutely* true? Not absolutely. I do not miss our ambivalence, her resentment, and her disappointment.
- Question #3: How do you feel when you have the thought "I miss her"? Miserable, remorseful, stupid, ashamed—which means my thoughts were what was torturing me, not her.
- Question #4: Who would I be without the thought "I miss her"? I would be my usual happy self.

Katie says to then turn the thought around—I miss her becomes "I miss me." I miss my normal, happy, sound-sleeping, wake-up-singing, healthy, driven, successful self.

These four questions and the turnaround can literally change your life. I have seen it work for many of my patients as well.

Lesson #3: The dance of relationships is both chemically and environmentally driven. Understanding the ingredients of the dance can help you be more effective both when the music starts and stops.

RULES FOR HIM, RULES FOR HER

Navigating Gender Differences in the Brain

"The genetic differences between the sexes are hundreds of times more significant than the differences between the races. You can't look at an fMRI of someone and say, 'That's an African-American brain, or a Caucasian brain.' But you can differentiate between a male and female brain."

—STEVEN B. JOHNSON

When Nicole threatened to leave Christopher, she said that their communication was like "a square peg in a round hole." They came to see me on the advice of a friend. Christopher was completely confused during their first session. He had no idea what she meant. Nicole said that she was tired of him not listening to her, always wanting to fix her, and not being present for her when she needed him. Bewildered, Chris argued that he loved her, spent hours listening to her, and just wanted to be helpful to her. I watched in pain as this couple played out a common gender war. I had once been victim of the same drama myself. When the love relationship I discussed in the last chapter ended, the woman used the same phrase as Nicole, "square peg in a round hole." Hearing it again, I felt a sharp pain sting my heart.

Male-female communication styles are radically different, brain-based, and hardwired. Many people think these differences are culturally defined, yet they appear very stable across almost all cultures studied and have their roots in the brain. Men and women evolved different brains over millions of years, because of necessity. Men chased down food for their families and provided protection. Women cared for the young and old, and provided a nurturing environment. Because of these different roles, the sexes evolved to process information differently, think in radically different ways, expect different things, and have different perceptions, beliefs, and behaviors. In the last forty years, as the gender roles in our society have been blurred, both males and females have become more and more confused as to what is normal behavior. We expect our partners to be able to read our minds and think as we think. Unfortunately, we are just not wired that way.

Knowing about the differences between men and women will help you be more effective in interacting with the opposite sex and help you navigate relationships without feeling hurt or rejected. In this chapter I will look at how male and female brains develop, the differences between the left and right hemispheres (which gives us clues to male-female differences), the differences in language and intuition, and the answers to commonly asked questions. I will also give males and females eight rules each for effectively dealing with each other's brains.

Different from Conception

The moment of conception determines gender. Males get an X chromosome from the mother and a Y chromosome from the father (XY), while females get X chromosomes from both Mom and Dad (XX). Inheriting a Y chromosome triggers two bursts of hormones, mostly testosterone, that change the brain and body. The first burst early in the womb differentiates the boy's brain from the girl's. From early in infancy, girl brains are more interested in smiling, communicating, people, and security; while boy brains are more interested in objects, actions, and competition. With higher levels of testosterone, the male areas of the brain are more developed, including the parietal lobes responsible for sense of direction, visualizing objects in three dimensions (good for catching a football), and mathematics. Without testosterone, the language centers of the brain are more developed, which is why girls like to talk and boys like to play catch. In addition, testosterone beefs up the area of the brain that is interested in sex. This area is twice as large in men as women. He really is more interested in sex.

The second burst of testosterone during puberty starts to turn a boy into a man. He now has twenty times the level of testosterone as most girls, his testicles descend, his voice deepens, his body becomes hairier, he has more well defined muscles, and he begins to grow like a beanstalk. His mind becomes highly focused on girls.

Since the female brain is exposed to little testosterone, her sexual development is very different. Her brain sends a message to the ovaries that it is time for the body to change. The ovaries begin to make much higher levels of estrogen and other hormones that cause the body to change in shape and size. They get the body ready for reproduction. Her breasts begin to grow, the nipples get larger and darker, curves begin to form as fat is deposited in the breasts and hips, and hip bones widen, which makes the waist look smaller. Her moods may begin to shift, she becomes much more interested in her appearance, and boys become the primary topic of discussion.

Two Minds in One Brain: Left Versus Right

The brain is divided into two sides or hemispheres, left and right, connected by three nerve fiber bundles. The largest nerve bundle is called the corpus collosum. Each hemisphere specializes in certain functions (although there is considerable overlap). We call one hemisphere dominant, usually the left side, because it is responsible for handedness and language, and the other side nondominant, usually the right side, because it houses other functions that are not as obvious. The left hemisphere is dominant in almost all right-handed people and about half of left-handed people. For this discussion I'll refer to the left hemisphere as the dominant hemisphere and the right hemisphere as the nondominant.

The left hemisphere has received bad press through the years—being thought of as the uptight, materialistic, controlling, unfeeling, and unemotional side of the brain. In fact, the left hemisphere is likely responsible for human success. It is analytical, logical, precise, detail oriented, and capable of conceiving and executing plans. The left hemisphere has more gray matter, which means a greater density of nerve-cell bodies. The left hemisphere is communicative and time sensitive, breaks down complicated patterns into component parts, and is heavily involved with planning. It is the happy, cheerful, more optimistic side of the brain. When there is good activity in the left-frontal lobe, people tend to be joyful and approach their lives with a positive zest for life. When the left hemisphere is hurt, people are often depressed, negative, and irritable. Sixty percent of people who have a left-frontal lobe stroke will develop a major depression within a year. A consistent brain-imaging finding in major depression is low activity in the left-front side of the brain.

The right hemisphere sees the big picture, or gestalt of situations. It allows us to see the forest, while the left hemisphere is responsible for the trees. The right hemisphere tends to be dreamier and more artistic. It helps us recognize familiar faces and is responsible for hunches and intuition. It also sees patterns at a glance. The right hemisphere has more white matter. Its neurons are connected from farther distances, which help it to draw on several different brain regions at a time. This allows the right hemisphere to come up with broad, multifaceted but vague concepts. The right side of the brain also allows people to know when problems are present and when they should be taken seriously. Unlike the left side, when the right hemisphere is hurt, people are often unduly happy, jocular, and in denial of any problems. People with right-sided strokes may develop "anosagnosia," failure to recognize disabling conditions. In the face of adversity, people with

right-hemisphere damage may appear unconcerned or even optimistic. The right hemisphere tends to be more negative, fearful,
anxious, mournful, and pessimistic. In looking at the differences
between the left and right hemispheres, it is as though there are
two individuals in one skull.

There are several practical applications of left-versus-right-
hemisphere research, such as where to stand in a conversation, or
what side of your partner's neck to kiss during lovemaking. Due to
how our brain processes information, if you stand to someone's
right side, you are processed more prominently in the left side of
his brain, the happier side of the brain. If you stand to someone's
left, you are processed more in the right side of his brain, the more
anxious and negative side of the brain. Some research indicates
that salespeople do significantly better when standing to someone's right. When you ask someone to marry you, interview for a
job, try to make a sale, or preach from the pulpit, stand to the
audience's right side, so that you are processed more in the happier left side of the brain; you may have a better chance of getting
what you want. In a similar way, if you kiss someone on the right
side of the body, she likely processes the kiss more intensely on the
left side of her brain, the happier side of the brain, and she is more
likely to want you to continue. If you kiss her on the left side of
the neck, she processes the kiss more intensely on the right side
and may feel more bugged or irritated by the kiss. Pay attention to
the reactions of others, depending on where you stand and how
you touch them. It may help.

Gender Brain Differences

Several of the gender differences in the brain correlate with the
left-right hemisphere differences. It has been reported that men
have more total brain cell numbers, even when corrected for their
increased total body weight. It has been estimated that men have
4 percent more neurons than women do. When I state this fact in
lectures, many women groan and complain. I then tell them about

a time when I was on the Carolyn Davidson radio show in Dallas, when she asked me, "Tell me Dr. Amen, why do men need one hundred grams more brain tissue to get the same things done as women?" It has been estimated that women have a greater number of cellular connections than men.

The corpus collosum, a large fiber band that connects the hemispheres, has been found in some studies to be larger in women, allowing greater access to both sides of the brain. Men tend to be more left sided in their approach to life, while women tend to use both sides more often then men. This is very important in helping us understand the differences between men and women. In language, men tend to be totally left sided, more detail oriented, and more directly to the point. Women, more often than not, use both sides, and tend to be more fluent, which may be why they have more to say. When husbands and wives get into arguments, their brain hemisphere activation styles often make problems worse. Because women use more words, men often become overwhelmed in disagreements and say insensitive things like "What's the point?" or "Specifically, what do you want?" These statements infuriate women and make communication even more of a struggle. They know the point, they just have more to say about it with the extra input from the right side.

The female hormone estrogen encourages brain cells to develop more connections within the brain and between the two hemispheres. Because of these extra connections, women are better at multitasking. They can talk on the phone while watching TV, cooking dinner, and checking their e-mail. Men, on the other hand, are more compartmentalized and do best when they do only one thing at time. When a man stops his car to read a map, he asks everyone else to be quiet and turns down the radio. This allows him to focus. His wife usually doesn't understand, as her brain can do, and actually enjoys doing, many things at a time.

Left-sided strokes, which affect the language area, tend to affect men more than women because many women have language on both sides of the brain. In one brain-imaging study, when

THE BRAIN IN LOVE • 79

asked to judge whether nonsense words rhymed, men used only the left side of the brain, where more than half of the women used both sides. In another study done at the University of Indiana, when listening to a novel being read, men used the left side of the brain, where women typically used both sides.

The limbic system, or emotional brain, tends to be larger in women. The limbic system is the emotional bonding center of the brain. Therefore, it is no accident that women in overwhelming numbers are the primary caretakers for children. Likewise, women are primary caretakers for the elderly; 70 percent of elderly people who need care get it from a woman. The larger limbic size makes bonding easier for women. Women tend to have more friends in life, they go to church more often than men, and they pray (bonding with God) more than men. Women have a larger nesting instinct than men. They have a greater biological need to have their houses in order. When a couple moves, it is generally the woman who feels unsettled until everything is put away. Women are usually primary caretakers for the home and take on the bulk of housework. With the larger limbic size comes a greater incidence of depression. After puberty, females are three times more likely to develop depression than males. Women attempt to kill themselves three times more often than men. Yet men, due to lack of bonding and use of more violent means, kill themselves three times more often than women.

In another brain study, the inferior parietal lobe was found to be larger in men, especially on the left side. This part of the brain estimates time, judges speed, visualizes objects in 3-D (such as catching a football thrown toward you), and solves math problems. Our direction sense is influenced by this part of the brain. Men are usually better at directions. Women, because of their greater access to the right hemisphere, are better at knowing when they are lost. Women are also better at knowing when problems are present. Psychiatrists see three times the number of women as men. Not because women have more problems, but because women are more aware that a problem exists and more willing to seek help. Women, more often than men, call marital therapists for help with a troubled marriage, and they bring their children to the child psychiatrist much more than fathers do. Seventy-five percent of the time, divorces are filed by women.

Recent research has shown that men may have a tendency to be more unrealistically optimistic. Unrealistic optimism is the belief that good things are more likely to happen to them than other people. If this way of thinking dominates a person's thoughts about life, it can have both positive and negative consequences. On the positive side, it is associated with better mental health as people are more hopeful and optimistic even in the most difficult times. However, this can prevent people from taking preventive measures for a negative outcome.

Sensory Skills

Women are better at reading facial expressions and noticing the feelings of others. Like most mammals, they are equipped with more sensitive sensory equipment. In a study by Barbara and Allan Pease, authors of *Why Men Don't Listen and Women Can't Read Maps,* they found that women were dramatically better at reading emotions than men. At a maternity hospital they collected a selection of ten-second clips of crying babies and asked the mothers to watch the clips with the sound turned down. This way the mother only had the visual information. Most mothers could quickly identify a range of emotions from hunger and pain to gas and tiredness. When the fathers took the same test, their performance was awful, less than 10 percent of fathers recognizing more than two emotions. Grandmothers also faired much better than the fathers, while grandfathers often did not recognize their own grandchildren. On the other hand men have shown greater blood pressure reactivity to sexually arousing pictures than did women.

Women also have better peripheral vision than men, which is why they catch their guys looking at attractive women, but hardly ever get caught when they look at hot guys. As hunters, guys have better long-distance tunnel vision. Men are also more adept at driving at night.

From birth, girls are dramatically more sensitive to touch, and as an adult, a woman's skin is at least ten times more sensitive than a man's. Women like and need to be touched more than men. In one study, a woman is four to six times more likely to touch another woman in a social situation than a man would another man. A woman's senses of taste and smell are also more sensitive than a man's. Men score higher on salty and bitter tastes, while women score higher on sweet and sugary tastes, which explains why there are more female chocoholics. Not only is a woman's sense of smell better than the average male, during times of ovulation she is better at picking up male pheromones that cannot be detected consciously. The superior sensory equipment of women allows them to pick up body language, thus dishonesty, much more than men.

She Is Thinking, Thinking, Thinking

Neuropsychologist Ruben Gur of the University of Pennsylvania used brain scans to show that when a man's brain is in a resting state, at least 70 percent of his brain is shut down. On the other hand, when women were resting, at least 90 percent of their brain was active, confirming that women are always thinking, thinking, thinking. The man wants the remote control and a little peace and quiet; she wants to talk.

In another study, when asked to think of nothing, men's brains were more active in the more primitive physical activity centers of the brain (cerebellum), women's brain were more active in the emotional and bonding centers of the brain (limbic system). Left to themselves, men will think about sex, their golf swing, or their jump shot; women will think about their spouse, children, or parents. One common complaint from women is that they do not feel connected to their partners. Men would do well to initiate conversations about children and parents.

As part of a national talk show on the differences between male and female brains that I did with gender expert Michael Gurian, author of the wonderfully insightful book *What Could He Be Thinking?*, I performed brain scans on Jennifer and Brad. They had the typical couple complaints. She wanted more time, talking, and attention. She wanted Brad to help around the house and be more present with the children. He wanted more peace, quiet, and sex. He wanted to be left alone for a half hour when he got home from work. Their struggles were clearly affecting their marriage. The brain scans showed clear differences. Jennifer's scan, as Ruben Gur's research would suggest, showed much higher levels of activity. Brad's scan had significantly lower levels of activity. Jennifer's worries and overthinking in the relationship were a product of a much more active brain, while Brad's need for rest came from his sleepy brain. A great couple activity to help balance both these states is physical exercise. For men, exercise wakes up the brain, while for women exercise enhances brain serotonin levels and calms the overactivity.

Sex Is Like Shooting Free Throws

When I counsel couples, I often say that sex is like shooting free throws, hitting a golf ball, learning how to throw a football or a curveball, or hitting a winning tennis serve. Men like sports analogies; women want their mates to pay attention. When boys learn to shoot free throws, or become skilled in any other sport, they practice over and over and over again. They repeatedly work on their technique. They spend hours at the free-throw line, figuring out what to do to improve their stats. Successful athletes have great coaches and they listen to them. They spend years perfecting their techniques and are not crushed by failure; rather they use it as an opportunity to learn. They film their performances to see how they can improve on technique and outcome.

For optimal satisfaction in the bedroom, women would do best to act like a good coach, recognizing that it may take your man some time to perfect his technique. Like a good coach, offer encouragement, praise, and advice. Meet him for practice on a regular basis and make it a fun experience that both of you want to revisit again and again.

Unfortunately, many women have trouble asking for what they want sexually. They may tell their man once or twice, often hesitantly, but then never again bring up their wants and desires. Many women I have counseled believe that if a man doesn't get it right the first time around, he probably doesn't really love, care, or want to please her. Training your lover to please you sexually is an important goal for your overall health. This is especially true for women, as their longevity is associated with pleasure, not frequency. Men are often slow to pick up the needs of their partner. They do not have the same access to reading social cues as women, so need more direct communication. Men are not born knowing how to please their partners; they need to be taught, over and over, like shooting free throws. They need good coaches who help them with technique and form. And you need to be satisfied; your life may depend on it.

The same principle applies to men. Ask what pleases your partner, listen, and practice. Tell her what you like, encourage her to talk about what she likes. Talk about what you do together that makes you think about her during the day when you are apart.

Nine Common Questions

As I was writing this book, the daytime television show *The View* did a program on the male brain. Elisabeth Hasselbeck, Star Jones, Meredith Vieira, and Joy Behar wanted to know why men and women were so different. My segment was on after talk-show-host Donny Deutsch, whom the girls had just roasted. They asked him how many times he had been married (which was twice), and why his marriages had failed (he didn't know), and whether or not he cheated on his wives (no), and what his problem was (he asked if they had a shrink's couch nearby).

Thankfully, they were easier on me. My interview covered nine questions, which are included below (these are some of the most common questions women ask about men):

1. "Is there really a difference between a man's and a woman's brain?"

Gathering my composure, I said, "Huge differences, and we can prove it." Here I pointed to a set of brain scans that we did of a couple at my clinic. "Typically, the woman's brain is very active. Thinking, thinking, thinking, especially in the emotional part of the brain. The man's brain, by comparison, is relatively quiet. A woman's brain is always working and a man's brain needs stimulation."

Meredith Vieira chimed in, "Seems like little is happening in the male brain."

2. "Let's get down to specifics. Why do men think about sex all the time?"

"The part of the brain responsive to sex hormones is two and a half times larger in men than women. Men are programmed to be more responsive to sexual feelings. Plus, with the lower activity in the brain, men are looking for excitement and stimulation. What is more stimulating to think about than sex?"

3. "You say there is a way you can tell which men have more testosterone than others. How?"

"According to University of Liverpool researcher John Manning, the size of your ring fingers and genitals are directly related to how much testosterone you received in the womb; the higher the testosterone level, the longer they are. In fact, looking at the length of ring fingers in comparison to index fingers will give an idea of the size of a man's penis. If the ring fingers are longer, it means that there were healthy testosterone levels; if they are the same size or smaller, it means that there were lowered levels. Women can estimate the length of a man's penis by saying, 'Show me your hands.' Those who have unusually long ring fingers (indicating very high testosterone levels) are at greater risk for autism, dyslexia, stuttering, and immune dysfunction. A large male member may not be all that great. On the other hand, a male with an unusually short ring finger is at higher risk for heart disease and infertility. Size matters, but it can go both ways."

4. "Why don't men need foreplay like women do?"

"Men are always ready for excitement. With the lower activity levels in our brain, and higher testosterone levels, it takes little to get us going. Like Harrison Ford's character Quinn Harris said in the movie *Six Days, Seven Nights,* 'All women have to do is show up.' We are always idling, waiting to be taken for a ride. Women, on the other hand, have so much going on in their brains that they need to be soothed, courted, and encouraged to be in the mood. They need a method to calm down their brains."

5. "Why don't men ask for directions when they are lost?"

"Men do not know that they are lost, even if they have passed the same gas station four times. Men have less access to the right hemisphere, which shows the big picture. The right hemisphere allows people to know when a problem exists. Since men have less access, they are often in the dark when something is wrong. It happens in many other situations besides driving, such as admitting that there is trouble in a relationship (75 percent of the time women are the ones who file for divorce or leave a relationship). Men are also singularly focused on finding their way. Admitting that you are lost is admitting failure, something men are very poor at doing."

6. "Why don't men gossip?"

"Men do gossip, just not as much as women. Men have language exclusively on the left side of the brain, while women have language on both sides. Women have more to say because more areas

of the brain are dedicated to language. Men are also more interested in sports than in emotional relationships. Sport talk shows
are very popular sources of gossip for men. In conversations, men
often get lost in the amount of words used by women. When it
comes to parenting, using fewer words is actually helpful. As a
child psychiatrist, I have seen that children usually listen to their
fathers more often than their mothers. In large part, it is because
men use fewer words than women and they are more serious about
consequences to negative behavior. Mothers want children to
understand, they want to be relational with them; fathers want
children to comply, like listening to the coach during a ball game."

7. "Why can't a man let go of the remote control?"

"With the lower overall brain function, a man needs more stimulation. As he flips through channels, he is often looking for something new, something different, something exciting. A woman is
often content watching a single program, with a relational story
line where characters show emotions. She likes to predict what
will happen next and needs the continuity to stay interested. The
remote control allows a man to surf the wild looking for fair game.
Men also have shorter attention spans than women. Men are diagnosed with attention-deficit/hyperactivity disorder (ADHD) five
times more than women."

8. "A recent survey came out and it said that men lie more than women, but that women are actually better liars. Why is that?"

"Given the lower brain function, especially in the front part of the
brain (also associated with the higher incidence of ADHD), men
tend to be more impulsive than women. They tend to have more
affairs and tend to say things without fully thinking it through and
often find themselves in hot water. Women also lie, but they get
caught less. One reason, as I mentioned earlier, is that women
have better access to the right hemisphere and thus read social

cues better than men. They notice the small things, like looking away or down or clearing your throat, that are typical in people who are not telling the truth. Since men tend to be in denial a lot of the time, they do not see when their partner is lying."

9. "Why don't men remember dates like birthdays and anniversaries?"

"Men are men. As such they are thinking about providing for their families, competing with other men, and sex. Opposed to women, who are thinking, thinking, thinking about the relationship, men are busy doing. This does not mean they love less, they just love like men. This often causes them to forget important dates, even though they try hard not to."

Rules for Men on Women and for Women on Men

Given the latest neuroscience research, here are eight rules to help men and women better understand and navigate the brain differences between the sexes.

RULES FOR MEN ON WOMEN

1. Recognize women are very different from you. Ask her what she needs to be happy and listen. Remember they leave the relationship 75 percent of the time.
2. Women typically need listening, not solutions (she is already competent).
3. Never ask a woman to get to the point.
4. She needs time, talking, and nonsexual touch for foreplay. (Her skin is ten times more sensitive. Find out how she likes to be touched. Her reality may be different from yours.)
5. Just because she catches you ogling another woman does not mean she is not ogling another guy.
6. On a long trip do the night driving; usually your night vision is better than hers.

7. She thinks it is sexy if you ask for directions. You win her heart by being willing to ask for help.

8. She has a keener sense of smell. Find out the smells she likes. Does she like it if you take a shower before bed, or does she like it if you don't? Find out.

RULES FOR WOMEN ON MEN

1. Recognize men are very different from you. Ask him what he needs to be happy.

2. He can do only one thing at a time. When you want to talk to him, wait until the game is over and ask for a specific time to talk.

3. Never try to get a man to admit to losing a fight.

4. If you want him to really listen, try to use fewer words.

5. He is programmed to compete and win. Make him think he wins a lot.

6. Just because you catch your guy ogling another woman does not mean you do not do it as well. It does not mean he finds you less attractive.

7. On a long trip do the day driving. Let him drive at night. His night vision is usually better. (And while you are sleeping, it won't bother you as much when he is lost!)

8. Many of his senses are not as keen as yours; tell him if odors or tastes bother you.

Lesson #4: Understanding the differences between the male and female brain helps prevent misunderstanding and pain and fosters communication and love.

LOOK CLOSELY

——

Brain Imaging Secrets to Enhance Your Love Life

> *"The brain is a tissue . . . it is a complicated, intricately woven tissue, like nothing else we know of in the universe, but it is composed of cells, as any tissue is . . . the connections that constitute the brain's woven feltwork can be mapped. In short, the brain can be studied, just as the kidney can."*
>
> —DAVID H. HUBEL (1981 Nobel Prize winner)

In the winter of 2005, *The New York Times Magazine* did a story on my brain-imaging work. In it the writer quoted me as saying if you date one of my daughters for more than four months, you have to get a brain scan. After the article appeared, I received several pieces of angry mail from people around the country saying that I used imaging to discriminate against people with mental illness. One woman wrote that she hoped my daughter married someone with bipolar disorder. Ouch.

It is true that if you date my children for more than four months, you get the opportunity of getting scanned. So far, everyone has taken me up on the offer. Inquisitive people want to know. I never use them to discriminate against people, rather as an opportunity to gather more information, like meeting someone's parents or taking them on vacation.

Brain imaging teaches us many important lessons regarding sexuality and relationships. In this chapter I will share six secrets I have learned from imaging to enhance your love life.

Secret #1:
There Is More to Love Than Most People Think

Looking at the brain has taught me that there is so much more to love than most people think. Biological brain influences are extremely important to how we feel, think, and act. Over the past sixteen years I have conducted brain-SPECT studies on more than three hundred couples who have had serious marital difficulties. I affectionately refer to this group as "the couples from hell study." I have been fascinated, saddened, and enlightened by these images. I look at love and relationships in a whole new way (as compatible and incompatible brain patterns). I have come to realize that many relationships work or struggle because of healthy brain activity or brain misfires, and they have less to do with character, free will, or desire than most people think. Many marriages or relationships are sabotaged by factors beyond conscious or even unconscious control. Sometimes targeted brain help can make all the difference between love and hate, staying together or divorce, effective problem solving or prolonged litigation.

Many people, especially some classically trained marital therapists, will see this idea as radical, premature, and even heretical. But how can we keep the brain out of the equation of love, sex, and relationships? Frankly, I know of no marital therapy system or school of thought that seriously looks at the brain function of couples who struggle. But I wonder how you can develop paradigms and "schools of thought" about how couples function (or don't function) without taking into account the organ that drives their behavior, namely the brain.

Brain-SPECT imaging is a powerful tool to evaluate brain function. It gives a sophisticated look into the function of living tissue. Simply put, the images give three pieces of information, showing areas of the brain that work well, areas of the brain that work too hard, and areas of the brain that do not work hard enough. This information then allows scientists to evaluate different brain systems.

One of the fundamental principles underlying our work at the Amen Clinics is that defined regions and circuits in the brain tend to perform certain tasks, and problems in these areas tend to give specific types of cognitive, behavioral, or emotional difficulties (see Lesson Two for a detailed look at brain systems). Balanced activity across the brain increases the chances for healthy behavior, while overactive or underactive areas of the brain can be involved in trouble.

Secret #2:
Whenever There Is Sexual or Relationship Trouble, Think About the Brain

The idea of scanning unhappy couples was born out of frustration and my internal critical voice. Bob and Betsy brought their two children to see me for school problems. As I worked with the kids, I came to believe that one of their major problems was the conflict between the parents. The chronic tension at home was having a negative impact on the children, causing anxiety, stress, and physical symptoms such as headaches, tummy aches, and problems concentrating. I suggested to the parents that they see me for marital counseling. They told me that they had seen four other therapists and it almost always made things worse for them. I was much younger then, more naive, and had the belief that I could help them. Maybe, I thought, they just hadn't seen anyone really good (through the years I have learned that thoughts like this one are usually a sign of immaturity or narcissism on my part and often a predictor of trouble).

In my office in our Northern California clinic, where I saw them, I have two navy blue leather couches. On Bob and Betsy's first visit, and every weekly appointment thereafter for the next nine months, they sat on opposite ends of each couch. That is a bad sign in marital therapy. After seeing this couple for three months, I started to hate them. Nothing I did with them seemed to make any difference. In my psychiatric training, my main supervisor said my biggest flaw was that I wanted patients to get better fast. I needed to be more patient. I dislike feeling ineffective. I thought to myself that Betsy had a PhD in grudge holding. She would go on and on, talk about the same things over and over, and be unable to let go of hurts from the past. Things from fifteen years ago still bothered her. I once thought that she would not only beat things to death, but also beat them into the afterlife as well. Betsy was married to a man I call "the sniper." He earned that nickname because he rarely seemed to be paying attention. Yet, whenever Betsy would settle down in her complaints, Bob would say things that were so evil, so nasty, just to get her going again. It seemed like he was purposefully revving her up. After six months of seeing this couple, I started to get physical stress symptoms on the day of their appointment. My stomach would hurt and my shoulders would get knotted. With no change in their marriage by the ninth month of therapy, all of us were feeling frustrated.

One day I was taking a shower getting ready to come to work and realized that they were on my schedule that morning. My stomach started to hurt. "Damn," I thought, "these people are in my shower with me. Today I am going to tell them to get divorced." I had actually been having that thought for several months. Research shows that it is better for children to be from divorced homes than from homes with chronic conflict. My problem with the thought about divorce, however, was that I grew up very Catholic. Not a little Catholic, but *a lot*! I attended Catholic school and was an altar boy for many years. We prayed the rosary every week on the way home from Grandpa's house. Even though I was no longer a practicing Catholic, I still lived with the Catholic voice in my head, commenting on my thoughts and actions. After I had the thought "Today I am going to tell them to get divorced," the internal Catholic voice yelled, "What! Because you are not a good enough therapist, you are going to tell these people to get divorced and damn their eternal souls to hell!" I just started to stare at the water faucet wondering, "How much therapy does this take to get over?" I got out of the shower, dried off, and reached for the telephone. I called my friend who owned the imaging center and said, "Hey, Jack, will you give me two scans for the price of one?"

He asked why.

I said, "I have this couple I have been seeing and have no idea how to help them. I am hoping to get some clues from their scans."

"Couple? You want to scan a couple," he said at first with disbelief and then curiosity. "How interesting! You know I have been married twice and I cannot figure it out. Maybe we could even start an Internet dating service and call it brainmatch.com."

When I presented the idea of the brain imaging, the couple was very interested. They were obviously aware that things were not getting better. Plus they would not have gotten divorced if I had suggested it. They wanted to be married. After all, I was the fifth marital therapist they saw.

Their scans literally changed their lives and mine as well. The woman's scan showed marked increased activity in a part of the brain called the anterior cingulate gyrus, which is the brain's gear shifter, allowing the brain to go from idea to idea and task to task. When it is overactive, people tend to get stuck on negative thoughts and behaviors, such as worrying or holding grudges. Just by random chance, if you believe in random chance, the night before her scan I had read an article in the *American Journal of Psychiatry* that reported increased activity in the anterior cingulate gyrus is calmed by Prozac. I put Betsy on Prozac. Bob's scan showed low activity in his prefrontal cortex when he performed a concentration task, a finding very consistent with attention-deficit/hyperactivity disorder (ADHD). At the time, in 1991, I was considered one of the experts in ADHD. I was very irritated that I had missed it in Bob. I put Bob on Ritalin. I told the couple to take a month off therapy to allow time for the medications to work (and for my stomach and the Catholic voices to settle down). When they came back a month later, for the first time they sat on the same couch. Betsy had her hand on Bob's leg, which is a good sign in marital therapy. Now, fifteen years later, they are still married, have another child, and get along better than ever. I call this better marriage through biochemistry.

The work with this couple caused me to rethink my work with all of the couples I saw. How many of their problems were learned behavior? How much was brain? How would I know unless I looked? Obviously, because of costs and availability of imaging, I couldn't scan everyone, but when I was struggling with complex cases, scans became invaluable. Over a decade ago, my friend and colleague psychologist Earl Henslin started sending many of the couples he saw to my clinic. He said of the first forty cases he sent to me, thirty-nine remained married years later.

Secret #3:
Something Completely Unexpected May Be Causing Trouble

Looking at the brain has taught me that there may be completely unexpected problems causing trouble for couples. The following story highlights this principle.

Dave and Bonnie saw a psychologist for marital therapy for three years. It was a frustrating endeavor. Try as they might to get closer, nothing seemed to work. The therapy sessions were filled with blaming, bickering, frequent explosions, and a general sense of unhappiness. The therapist, who was very experienced, tried and tried, yet nothing worked. After considerable thought the doctor decided to give the couple an F in marital therapy. She told Dave and Bonnie that in her opinion it was time for divorce. When the couple protested, as they had spent years of effort and over $25,000 trying to get better, the therapist said there was one more option. She told them about the work at Amen Clinics, where some of her most difficult clients had been helped.

After an evaluation we performed brain-SPECT scans on the couple. Dave's brain scans looked shriveled and full of holes, the same pattern that we see in drug or alcohol abusers. The scan was odd, because in his history Dave said he didn't drink and never used drugs. To make sure, in front of Bonnie, I asked Dave if he was drinking heavily or using drugs. He said no, and repeated that he didn't drink and never used drugs.

I turned to Bonnie for more information, knowing that alcoholics are often in denial and drug abusers often lie. She said, "He is right. He doesn't drink and as far as I know has never used drugs. That is not his problem, Dr. Amen. He is just an asshole."

I chuckled at her comment. But internally my mind started to race. If he was not a drinker or drug abuser, then why did he have such a toxic-looking brain? I went through the different potential medical causes in my head: brain infections, near-drowning episode, hypothyroidism, anemia, and environmental toxins. My friend psychiatrist Harold Bursztajn, co-director of the Psychiatry and Law Program at Harvard, often says that scans are usually not the answer, they teach you to ask better questions. My next question to Dave was, "Where do you work?"

He replied, "I work in a furniture factory."

"What do you do there?" I asked.

"Finish furniture."

"Is there good ventilation in the room?" Oh my goodness, I thought, Dave has a drug-affected brain from the solvents he was using at work; even though he has never willfully used drugs, they are eating away his brain. Dave just thinks he is being a good provider to his family, while his brain is being poisoned.

"No," Dave said, "it is often hot and reeks with fumes."

"Do you wear a mask?" I asked.

"No, they tell me I should but I don't think it is important."

"Ouch," I said. "You really should."

My next question was to Bonnie: "When did he start becoming an asshole?"

She thought for a moment. "We were not always unhappy. We have been married for fifteen years. It just seems that the last eight were hard. The first years were great. He was so different."

Then Bonnie had a look of "Aha" wash over her face. "Dave started to work at the furniture factory eight years ago. Do you think his personality change can be from his job?"

"You bet," I answered. "Something is eating away his brain, and eating away his ability to be kind, thoughtful, empathic, and to love you."

Working with this couple has been an amazing lesson in the brain-love connection. I took Dave out of work for six months and would only allow him to return to a nontoxic job at the plant. Bonnie developed empathy for her husband, who before in her mind was just an asshole but who had become someone in need of help and understanding. When behavior does not make sense, it is important to consider brain health issues as a potential cause of the trouble. When people have issues concerning their behavior, the brain is an important place to look. Research done on brain-injured patients revealed that over one-third of them suffer from depression. In addition, of those people suffering from depression, three-quarters also had anxiety issues and exhibited aggressive behavior.

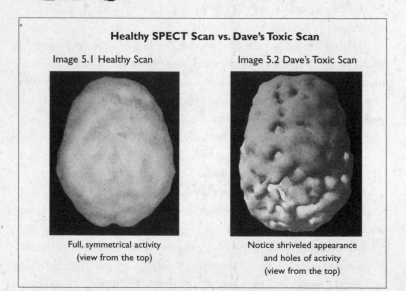

Healthy SPECT Scan vs. Dave's Toxic Scan

Image 5.1 Healthy Scan

Image 5.2 Dave's Toxic Scan

Full, symmetrical activity
(view from the top)

Notice shriveled appearance
and holes of activity
(view from the top)

Secret #4:
Think About Scanning Potential Partners or Taking Their Brain Science History

When I was single, whenever I dated someone new, I tacitly took a brain science history. I wanted to know as much about how her brain worked as I wanted to know how her body felt next to mine. If I thought the relationship had the potential of going further, I asked her to get a scan. Most women had no problem with the idea. In fact, when women really understood my work, most were curious and wanted to know more about themselves. The idea of scanning someone was never to rule them in or out; it was to gain a better understanding of the issues that might face us in a serious relationship.

One of my close friends, Will, had been dating a new woman he met on Match.com. He really liked her and felt they had great chemistry but he was concerned about a number of issues. She tended to run late, was disorganized, lived on the edge by riding motorcycles, and was prone to starting fights between them. As much as he liked her, she was not easy for him to be around. She suspected herself that she had attention-deficit disorder. Her scan showed very low activity in the PFC when she tried to concentrate, which is a common scan finding of ADD. The deeper understanding of his potential mate allowed Will to make a more informed decision about what he wanted to do with the relationship.

Another friend, Katie, started dating a man, Ben, she had met through her professional circle of colleagues. Again, the chemistry was powerful, but she was concerned with his tendency to argue and be oppositional. It never felt easy being with him. She felt on guard a lot and she was often chastised by her new boyfriend for what seemed to her like minor offenses. After first refusing to get scanned, he later agreed. His scan showed excessive activity in his anterior cingulate gyrus. His brain tended to get stuck in the loop of negative thoughts and behaviors. Recognizing the problem after seeing his scan, and admitting it had been a problem in other relationships, Ben took a supplement (5-HTP) to calm this part of the brain. Subsequently, the couple got along much better.

Since most people do not have access to scans, taking a brain science history can be of great value. I'll discuss this more in the next chapter.

Secret #5:
Brain Health Becomes Part of the Relationship

My imaging work has led me to want a better brain. I have personally been scanned ten times over fifteen years. As people age, the brain usually becomes less and less active. In my case, my brain has improved over time, because I have taken personal brain health seriously. In my relationships, whether with friends or potential partners, eventually everyone ends up being scanned. The imaging work generates intense interest. Most people want to know about themselves and how their own brain works. If one of my friends gets scanned, brain envy starts to invade his or her consciousness. She wants a better brain. She becomes more likely to eat in healthy ways, exercise, take supplements, wear her seat belt, and avoid too much alcohol or caffeine. As brain health enters her consciousness, it is easier for us to be together. We share a common goal, brain health. Ordering at restaurants becomes a collaborative effort and spending time together turns into more walks and word games than getting drinks at a bar.

Secret #6:
Scans Generate New Ideas

Scans generate new ideas and new ways of thinking about relationships. I recently saw a troubled couple. The husband was bipolar and his wife had gotten to the point where she couldn't stand his behavior anymore. Sex was one of the areas of discontent. She came in for the first time on a joint visit to add some information about how he was doing. Within one minute of being with her, I felt very uncomfortable. Her tone of voice had frequencies in it that made me uneasy. I thought it needed to be recorded so we could study the brain reactions of others. Years ago, my colleague and coworker Dr. Leonti Thompson saw a number of men who had killed their infants in a fit of rage. His research brought up several cases where tests had shown that the frequencies of infant cries triggered seizurelike activity in the temporal lobes, arousing feelings of anger. If the wife and husband in this case were to be scanned when the wife was talking to him, I wondered if we would see his temporal lobes fire abnormally in response to her irritating frequency.

I once scanned a man who lit his ex-wife's house on fire at Christmastime. He was on trial for arson. His initial scan showed mild trouble in his left temporal lobe (an area of the brain associated with violence). Then, I scanned him while he listened to a message from his ex-wife. During this scan his mild left temporal lobe abnormality became severe, providing us with evidence that her voice triggered a negative reaction in his brain. Without the scans I would have never been able to see this connection.

The following story was contributed by my friend Barbara Wilson, a neurologist and pain specialist in Austin, Texas. The case illustrates how brain imaging and changes in brain function alter behavior, which may be positive or negative.

"I had a very nice man who had a facial tic. It was just on one side. He smoked a lot of pot because it made his headaches not as bad. I initially trained as a movement-disorder doc—and one-sided tics must be scanned for tumors, even if the rest of the physical exam is okay. I got into trouble for ordering the MRI—the guy was in an HMO and they figured he had tics for at least three years so no immediate testing was necessary—but I stuck to my training and ordered the scan. The patient had a large frontal lobe tumor with a little spot of calcified brain. In neurology that little calcified spot upsets one greatly because it indicates that the tumor probably started out as benign, which would have been cured if it had been taken out earlier.

The really interesting part of the story: His girlfriend and I had a long talk. She spoke of how much fun he was—not inhibited like other guys. He was such a "unique spirit—free, wildly sexual, and unreasonably happy with no rigid boundaries like most guys" she knew. . . . Basically she was in love with the personality of someone without much of a frontal lobe. The girlfriend was devastated that this special wonderful man was diagnosed with a severely life-shortening tumor. Ironically, if he had not had the tumor, his personality would not have been attractive to the girlfriend. Consistent with the effects of some brain lesions; the guy wasn't that bothered by his impending death. Actually it kind of clarified things: He could just spend all his money on drugs since he didn't need to worry about long-term wealth accumulation. I worried more about the girlfriend, destined to go through life looking for another guy with a frontal lobe tumor.

This story illustrates how our sexual lives are tied to brain function. Whether we are outgoing, uninhibited, uptight, or afraid has to do with the moment-by-moment function of the brain. Looking closely at brain function opens a new world of possibilities for explanations about behavior and potential treatment enhancers.

Lesson #5: Looking at the brain can enhance your love life.

USE YOUR BRAIN BEFORE YOU GIVE AWAY YOUR HEART

Warning Signs of Trouble

"Know when to walk away,
And know when to run."

—"The Gambler," KENNY ROGERS

Have you ever made stupid decisions about love? I mean really inane, insane, unbelievably stupid decisions. I have, and some of those decisions have turned out to be very painful. I am never someone who feels as though I have it all together, but in the past I have felt as though I was smarter than I acted regarding some of my romantic decisions. And I have been curious about these less-than-stellar decisions for a long time. Can neuroscience help us understand the decisions we make about love? Can it help us know when to hold someone, when to walk away, and when to run? Decisions about love can help extend your life; as a group, people who are happily married live longer. Decisions about love can also ruin your life, through emotional pain and bankruptcy, and, in some cases, end it through suicide or murder. Decisions about love have potential life and death implications.

Warning Signs Around the Oasis of Love

When we fall in love, many people experience what I call the Oasis Effect. Coming out of the desert of being alone and longing to be in a relationship that helps them feel more complete (as humans we are wired to be intimately connected to others), many people find that new love feels like an oasis of beauty and nourishment. When we fall in love, there is a large release of the bonding hormone oxytocin. This chemical has been found to increase our sense of trust, even in situations where perhaps we should be more cautious. Similar to the dusty, dirty, lonely, and thirsty desert traveler who is euphoric upon finding the oasis, falling in love feels unlike anything else, exciting, fulfilling, and satisfying. When first coming out of the desert, travelers are often so ecstatic to find the waters that they fail to see anything else. Their anxious state of euphoria causes them to overlook or ignore warning signs around the oasis, such as diseased animals. Those who have drunk from those same waters were sickened, evidence of the poison along the edges of the oasis. So, too, in love, when we come out of the desert of aloneness, we are often so happy, with love chemicals coursing through our brain, that we fail to see the trouble before our eyes, the metaphorical dead animals around the oasis of love. Here are two examples.

Eric

Eric and Becky were in a tumultuous marriage for ten years. Eric had thought about getting divorced for the last five years they were together. They had seen eight different marital therapists without much help. One day, after going to a friend's twenty-fifth wedding anniversary, Eric decided he had had enough and saw an attorney. There was no way he wanted to stay in a marriage filled with distance and pain for fifteen more years. Even though the divorce was painful, Eric knew it was the right decision. In dating Becky, Eric had ignored many warning signs of pending trouble. Nowhere in school are people taught to notice problematic signs in a relationship. Becky had been severely sexually abused by an alcoholic father. When dating Eric, she failed to tell him of the tension, turmoil, and abuse at home. Without psychological help to process the pain of an abusive past, intimate relationships for an abused person are usually impaired; even with help, it can be a long road. In dating again, Eric vowed it would be different next time and he would make better choices for himself. Yet, his next two relationships were filled with turmoil and pain, prompting Eric to come and see me.

In describing his relationship with Becky, Eric said it was like living in a desert. Becky was not only volatile, she was almost always sexually withholding. Eric said his sexual advances were rebuffed nine times out of ten and he lived in a chronic state of frustration.

His first relationship after his divorce was with a beautiful, smart, highly sexualized model, Monica. For the next eight months he felt happier than he had ever been, so he and Monica moved in together. Then the poison in the oasis started to make Eric sick. Eric and Monica started to fight. Monica spent money after agreeing not to, hiding it from Eric. Sex, which was free flowing in the beginning, was now being withheld. And Eric started to catch Monica in a series of lies. Even though there were several significant signs of trouble before they moved in together, the bliss of the sex and the hope of finding a lasting, healthy love blinded Eric to the reality of the situation. He failed to see the warning signs:

Monica had a young child who did not live with her, usually a sign of trouble.

One of Monica's friends told Eric that she lied a lot and could not be trusted.

Monica's oldest child told Eric that she had gone through six of her mother's relationships and breakups, and that the last time she didn't come home in order to avoid the fights.

Eric's children always felt very uncomfortable around Monica.

In addition, Monica had posed nude for a famous photographer. Eric was fascinated by her ability to do this work without embarrassment. One of the nude photos hung in their bedroom. However, Monica wanted to put her second nude photo in the living room. When Eric said no way (they both had children!), Monica had a huge fit.

Sex and hope for lasting love can cause blindness, even in very bright people.

After his breakup with Monica, Eric decided to take time off to discover what he really wanted in a relationship. He felt emotionally wrung out and had very good reason not to trust his own judgment. Then, through the Internet, Kate came into his life. He felt she was different. She was sweet, present, loving, and seemingly sincere. Even though the sex was routine, it was ever present, and she seemed emotionally healthier than Monica. It felt like another oasis, where the water was cleaner and more refreshing. But the warning signs were again ignored.

Both of Kate's parents struggled with alcoholism.

Kate resented her father, whom she said loved her brothers more than her. This father-daughter dynamic usually causes women to resent other men they start to trust.

Kate had left two husbands, blaming them for the problems in the relationship.

She said upfront that she cheated the IRS and never reported her cash income.

The first few months of the relationship were great; then Eric and Kate went through a series of breakups and getting back together, which was always Kate's idea. When things didn't go her way, she left again.

Now the warning signs became flashing-red-neon signs. Still Eric ignored them:

Kate started pushing Eric to be more committed than he felt ready to be.

Kate never once said she was sorry or asked for forgiveness during any of their disagreements. It was always Eric's fault.

Kate never seemed satisfied or happy with Eric. When they finally broke up for good, Eric felt as though he had fallen off a cliff. He was in such emotional pain that he felt as though his skin was being ripped from his body without any anesthesia. He couldn't sleep for six months; he experienced crushing chest pain;

felt obsessed with the thought of Kate; had suicidal thoughts for the first time in his life; and despaired of ever having a healthy relationship.

That's where I came in for Eric. He needed to do a better job at seeing those dead animals around the oasis of love. Make no mistake; Eric was a big part of his own trouble. He was drawn to women whom, like his father, were distant and disapproving. Unconsciously, he found these women the most exciting. He tended to avoid women who were happy with him. His anxiety and actions subtly encouraged these women to take advantage of him, such as providing significant financial support without clear guidelines on what he expected in return. He also tended to set them up to be perceived as critical like his father by being excessively hurt by their criticisms rather than being able to hear them and try to improve. The work in therapy was about seeing his role in the demise of these relationships and helping him see the warning signs around any potential new love so he could be more cautious when he gave his heart away.

Jennifer

Can you see the warning signs?

Through the Internet Jennifer met John for dinner on the day after Christmas. He was single, had never married, was forty-six years old, and had a history of short-term relationships. When he met Jennifer, he just melted. He said he could not even talk in her presence and did not want to be apart from her. That evening he invited Jennifer to go to Paris with him the next day, which she did not do. By New Year's Eve they were talking up to eight hours a day, totally immersed in each other. From the moment he came back from Paris, ten days later, they were never apart for the next two months. Jennifer was totally taken with John. He was handsome, successful, highly intelligent, young minded, funny, charming, and independent. And, very important to Jennifer, he had the name John. For years, she was convinced that John was the name of her true soul mate. When they met, both of them dove right in without holding back. Jennifer loved that he wasn't afraid to love her. Little did she know about the pain that awaited her.

John was the most generous man Jennifer had ever met. If he gave her flowers, he brought four bouquets, not just one. He never went shopping for himself without thinking of her and buying her something. He was abundantly generous with his words and affection. He constantly talked about how much he loved her, how much he loved holding her, retelling the story of their meeting over and over again and how special and new this "feeling" was. He asked her to marry him within two weeks. He told her that she was the woman he had been waiting for. He had been waiting for "this feeling" and now knew why he had never married. Whatever was good, he'd reinforce by talking about it all the time. They talked for hours about how great their relationship was.

After two months, Jennifer decided she needed some space and would not sleep over at his house. John became irate. The relationship quickly started to unravel. He liked Jennifer when she was a sweet, soft, loving girl, but when she acted like the assertive woman she really was, he became angry and controlling. Their fights about their troubles went on and on. He had a challenged relationship with his own depressed mother, who often did not talk to him, which may have been why he needed to talk about problems for a long time. Initially, the relationship was so thrilling that Jennifer jumped in the water without hesitation. When the warning signs flared, she ignored them because she did not want to lose the hope of this new love.

Jennifer started to realize that everything had to be on his turf and his way or there was a huge power struggle. He was overly sensitive to her every move, gesture, tone, and mode of communication. If there was something Jennifer did that he didn't like, she would hear about it for hours . . . basically until she recognized the error in her ways and repented. One night, after another fight, she went home. A few hours later John came over, violently banging on the door until Jennifer opened it. Then he got furious at her because she wasn't excited to see him. They went into therapy, but after a few sessions the therapist saw it as hopeless.

The hope of love often blinds us to the reality of the situation. Losing hope is painful. Living in a mismatched relationship is usually much more painful than being alone.

Look Before You Drink

Before you give away your heart, use your brain. Notice the warning signs and any evidence that there will be potential trouble at the oasis of love. Never expect perfection, as there are very few perfect people in the world. But look for toxic symptoms. All of us have some trouble in our past; but how much and how severe is a score worth noticing. Here is a list of questions about potential new relationships to help you see any warnings.

- Do you often feel in trouble when you are with the other person?
- Is he or she disappointed in you a lot?
- Are you pushed to go faster than you are comfortable?
- Are your friends or family concerned about the relationship?
- Do you have a nagging, internal voice saying that there is trouble brewing?
- Are you ambivalent?
- Is there a past history of many short relationships? This indicates trouble committing or the constant need for someone new and exciting in his/her life.
- Is his/her behavior inconsistent?
- Does he/she blame everyone else for his/her problems?
- Does he/she have trouble telling the truth?
- Does he/she never apologize?
- Do his/her own children say negative things about him/her?
- Do your kids feel uncomfortable around him/her?
- Is there any drug abuse?
- Does he/she cheat the IRS?
- Are there temper problems?
- Does he/she act in a demeaning or belittling manner?
- Is there evidence of alcohol abuse?

- Do you lose yourself in his/her presence?
- Do you find yourself trying to mold yourself more and more to make him/her happy?
- Does he/she come forth with interest and then suddenly withdraw and become unavailable?

Taking a Brain Science History

When evaluating whether or not a person is a suitable partner for you, assess how his or her brain and psyche work. Can you really do this without a scan? Yes. Through the questions we ask ahead of time at the Amen Clinics, our historians are uncannily accurate in predicting what the brain scans of our patients are likely to look like. We assess brain function by asking the right questions. You can, too. I believe it is essential in dating to obtain extensive information on a potential mate before you give away your heart. Think about using your head to protect your heart.

What kind of questions should you ask? What follows is the outline of the major sections of the Amen Clinic Adult Intake Questionnaire that we use to take the histories of patients who come to our office for help. I will give you tips on how to gather this information and how to ask insightful questions. None of these questions need to feel intrusive. Do not ask them in rapid-fire succession, but rather weave them into the give-and-take of many conversations. Ask them over the first month of dating someone new. Write down their answers in private to obtain a detailed history and fill in the gaps as needed. Anyone who objects to this type of probing likely has something to hide.

ACI Adult Intake Topics

I. Why Are You Here?

In this section we ask people why they came to the clinic. What is the problem? For daters, you need to know what the real purpose of the date is. What does the person really want? This is a critical question to answer. Is the person looking for a one-night stand, a friend, a long-term relationship? Do his or her goals match your desires?

2. What Have You Done in the Past to Fix the Problem?

In this section of the questionnaire we ask patients about past attempts to fix the problem for which they are seeking help. For daters, you want to learn about their past dating experiences. How have they attempted to meet people? What has worked and what hasn't?

3. Medical History

Here we ask about a patient's physical health. We ask about general health, current and past medications, surgeries or hospitalizations, past head injuries, sexually transmitted diseases, and allergies. In dating, you want to know the specifics about the health of a person. A good way to ask these questions is by starting off with your own health history and then asking about theirs. The health of someone's brain usually influences the potential health of their relationships. Be curious. Past head injuries may help explain impulsive or erratic behavior. None of the information gathered needs to be the final decision on ruling someone in or out, but it can help you make a more informed decision in the context of the whole relationship.

4. Past Psychiatric History

Has the person ever been diagnosed or treated with a psychiatric illness, such as ADD, anxiety, or depression? According to the National Institutes of Mental Health, 49 percent of the United States population at some point in their lives will meet the criteria for a mental disorder. It is almost more normal to have a problem than to not have a problem. I never think this should rule someone in or out. But you need to know. If someone has a condition, you absolutely want to know as much as possible about it.

5. Current Life Stresses

In this section we ask about the current stresses in a person's life, such as financial, health, work, or family issues. It helps us to understand the whole person and perhaps why they are having problems at this time. The current stresses in a person's life are something worth knowing for potential daters. What is stressful or difficult in the person's life at the moment? A friend of mine once dated a woman who had serious financial problems. She was still being supported by an ex-boyfriend who continued to have sex with her as an exchange. It was important information for my friend to know. This can also be a place to ask potential partners about financial stability. Even though this is an uncomfortable topic, it is critical nonetheless. Financial issues are one of the most common sources of stress for couples. Truly understanding a person's attitude and behavior toward money can tell you a lot of information about a person's judgment and how he or she deals with responsibilities.

6. Sleep Behavior

Sleep issues are explored in this section, including nightmares, recurrent dreams, insomnia, and any current problems in getting up or going to bed. In sexual relationships sleep is a very important issue. Is the person a light sleeper, so any snoring might bother him or her? Is there chronic insomnia, so partners cannot sleep together, which is a very common problem?

7. School History

We want to know all about school. It is such an important part of someone's life. Here are some questions to ask. What did you like and not like about school? What did your teachers say about you? What were your best and worst subjects? What were your biggest frustrations and joys in school? How far in school did you go? Do you have further educational goals?

8. Employment History

Likewise, we want to know all about a person's work history. Work says a lot about a person. Some questions to consider asking include: What were your favorite jobs? Worst jobs? What would past employers say about you? What is your dream job? What are your work goals?

9. Legal History

This section explores any legal or criminal issues. For a potential partner it is a good idea to know if there are any ghosts in the legal past, such as bankruptcies, divorce proceedings, child-custody issues, arrests, criminal charges, and convictions. Even though it seems obvious to ask, sometimes we are blinded by the light of the oasis and fail to do so. If you are in doubt, check their background, which can be done easily on the Internet.

10. Sexual History

This is often an uncomfortable subject, even for psychiatrists. But it is essential to understanding a person's life and psyche. It is crucial to know the sexual history of your potential mate. Here are some potential questions to consider. Age of first sexual experience? Number of sexual partners? Any history of sexually transmitted disease? Any abortions? Any history of molestation, sexual abuse, or rape? Any current sexual problems? Any fetishes or particular behaviors I should know about? What is your attitude toward pornography?

11. Drug and Alcohol History

This is a critical part of our intake process. Even with us, many patients lie about their drug and alcohol usage. Of all the issues that damage relationships, drug and alcohol abuse are near the top of the list. Moderate to heavy use of drugs or alcohol likely predicts trouble in the future because they damage brain function. Whatever damages brain function damages life function as well. Ask candid questions, listen to the answers, and observe the person's behavior when you are together. Questions we ask about drug and alcohol usage include: How much alcohol do you drink? What other drugs have you used? What is your current usage? Has anyone ever been concerned about your drug or alcohol usage? Have you ever felt guilty about your drug or alcohol use? Have you ever felt annoyed when someone talked to you about your drug or alcohol use? Have you ever used drugs or alcohol first thing in the morning?

12. Significant Developmental Events

In this section we want to know about the events that help shape a person's character. Here we want to know about marriages, separations, divorces, deaths, traumatic events, losses, abuse, and also biggest joys, greatest moments, and successes. Listen to the events that highlight your potential mate's past.

13. History of Past Marriages and Romantic Relationships

This section helps us understand how people connect to past intimate lovers and partners. It is important to know about a person's history in relationships. Do they blame others for all the problems or have they learned from their own mistakes? What have been the issues, patterns, joys, and sorrows in past relationships? The best predictor of future behavior is past behavior. How people have been in the past is likely how they will be. This is not always true, but true enough that you want to know past patterns.

14. Family History (Parents, Siblings, and Children)

Family history is one of the most important sections of our intake process. Many brain problems, such as ADD, depression, alcoholism, or bipolar disorder, are genetic and tend to run in families. Understanding a person's family usually gives you good insight into their own character and vulnerabilities. Listen to family stories. Specifically, try to understand the relationship between a potential partner and his or her mother and father.

How did he or she get along with his or her mother (which is usually the primary bonding relationship)? This relationship usually sets the tone for all other relationships for males and females. A positive maternal-child bond helps the brain feel more settled as it develops; a strained maternal-child bond causes stress hormones to disrupt development. How did he or she get along with his or her father (another crucial relationship)? What about the relationship with siblings, and children? Meet his/her family and watch the interactions between them.

15. Spirituality

It is unusual for many psychiatrists to ask about a person's relationship with God or his sense of meaning and purpose in life. Yet, I feel these are very important topics to discuss. This is certainly a critical issue in relationships. When people are matched in their sense of meaning and purpose, with their beliefs in God or a higher power, their relationships tend to be happier and more mutual. Ask your potential mate about current beliefs in God and religion. What was he or she taught growing up? What does he or she believe are the influences of these beliefs on other aspects of life? What is his or her overall sense of meaning and purpose in life? Why does life matter? These questions often stimulate great discussions. If not, that is more information for you to have in evaluating the relationship.

BrainMatch.com and Internet Dating

Internet dating is the rage, and rightly so. Bars are definitely not the best place to meet someone, especially if you are into brain health. Workplace romances are often trouble from the start. And friends and family can only set you up so much. The Internet works because you are meeting people who want to be met—people who are ready, willing, and available (if they are not lying). People on dating sites help you sift through important information. They tell you what they like, don't like, their habits and interests. You can see how they write, how they think, and how creative or not they are. I have many friends and patients who have made great use of Internet dating.

Of course, there are dangers to Internet dating as well. One of my eighteen-year-old patients from Northern California met a man from West Virginia on the Web. They talked for hours, got engaged, and he sent her a train ticket to be with him. When she told me all of this in a therapy session, the father in me freaked, and the psychiatrist part of me called a family meeting. I had the mother and father come into our session and we discussed the pros and cons of the pending move. When the girl broke off the trip and the engagement, the man threatened to kill her. We later found out he had just been released from jail. Children, teenagers, and some young adults need to be protected from the seedy side of the Internet. As sex and gambling can become addictive, so can Internet dating. I have known people who spend hours a day searching for the perfect date. They can't stop and it affects every aspect of their lives, including their jobs and relationships with their children.

From a neuroscience perspective, here are a few tips for safer Internet dating:

1. Really look at the photos posted. If they contain pictures of motorcycles, fast cars, or skydiving, it is likely the person is excitement seeking (perhaps low in PFC activity). If more

than one of the pictures has alcohol in it, be concerned about drinking.

2. Look for the truth. Many people lie on Internet dating. If they lie in that situation, they are likely to lie in others. Lying breeds mistrust. If a person lies about his age, income, desire for children, be concerned. Lying is a deal-breaker.

3. Talk to them on the telephone and by e-mail at least three times. Be patient. Get to know someone before you let them know where you live. Remember the movie *Fatal Attraction*?

4. As in all dating, be careful about moving too fast. Dopamine chemicals are being pumped out fast and furious in the first few months of a relationship.

Falling in love can be so powerful that we ignore the potential pitfalls. Use your head and your heart when you fall in love. When emerging from the desert, be sure to take care of yourself. Pack your own canteen full of water so that you are less dependent on the oasis. Pay attention and take your time before drinking the water.

Lesson #6: Notice the warning signs in a potential new relation-ship—use your brain and your heart when you fall in love.

THE BRAIN IS A SNEAKY ORGAN
___

Addictions, Weird Sex, Fiends, and Fetishes

"The brain is a sneaky organ."

—JERRY SEINFELD

Does your brain play tricks on you? Surprise you? Torture you? Appall you? Mine does. Even at the most inappropriate times. I can be sitting in church, trying my best to be thoughtful, prayerful, spiritual, and close to God, and then the next moment my brain notices the rear end of the young woman in front of me.

"Stop it," my superego yells.

"Oh, please, just one more quick look?" My brain pleads.

"No, you are in church."

Or, I can be out to dinner with my sweetheart, having spent the day planning to make it a special evening for us, and then ruin it all by noticing for a half second too long the young, bouncy waitress with the ample cleavage walking by the table. Poof, the evening is ruined. "What's the matter with me?" I think, "I did not want the waitress, I wanted my sweetheart." Also, I can be driving in traffic and wonder what it would be like to drive into oncoming traffic or slam into the car next to me. Ouch! Or, I can be walking in a grocery store, notice someone carrying an armload of boxes, and wonder what it would be like to tickle the person. Yuk? The list goes on and on from suicidal thoughts ("What would they think if I jumped off the balcony and splattered myself on the ground?"), to homicidal thoughts ("How would it feel to shoot that person in the face?"), to strange thoughts ("I wonder if it would be erotic to watch anteaters having sex?"). Before you think I am abnormal or just plain sick, I have been listening to these "sneaky thoughts" from my patients for more than twenty-five years. We all have them. They just sneak up on us without planning. I was walking recently with a friend, one of the sweetest, most thoughtful women I know. She told me about a time when one of her daughter's friends was being irritating and she had the thought of pouring a jug of milk over her head. Of course, she didn't do it, but the thought ran through her brain nonetheless.

What causes our brain to have these bizarre, silly, or unhelpful thoughts? What causes our brain to be sneaky? The limbic or emotional brain is always generating possibilities, novelties, and interesting hypotheses. Like dream states while awake, the brain is constantly churning, imagining, and playing. Thankfully, we have an area of the brain called the prefrontal cortex (PFC), which inhibits these sneaky thoughts and prevents us from saying them or acting upon them. When this part of the brain works right, it can laugh at or dismiss these sneaky thoughts. When there is damage or disease to this part of the brain, these hurtful, embarrassing thoughts surface in our behavior.

I was once at a conference with a close friend of mine, Jillian. She had experienced a car accident several years earlier that hurt her PFC. She had a reputation for saying exactly what was on her mind without filtering its content. Two obese women sitting in front of us at the conference were engaged in a spirited conversation about their weight. One woman said to the other, "I don't know why I am so fat, I eat like a bird."

Jillian looked at me and said loud enough for everyone around us to hear, "Yeah, like a condor."

I looked at her in total embarrassment. Horrified, Jillian put her hand to her mouth and said, "Oh my God, did that thought get out of my mouth?"

Yes, I nodded.

"I'm so sorry," she said as the women moved away from us.

The brain is a sneaky organ. From minor gaffs, to embarrassing moments, to a lifetime of trouble, the brain is in the center of our behavior.

PFC Damage

Damage to the PFC can happen in a number of different ways, such as through a head injury, some form of toxic exposure, or later in life through diseases of aging, such as dementia. The most common cause of dementia that is associated with difficult behavior is called frontal temporal lobe dementia (FTD). People with this type of dementia are more likely to act like Jillian. In a study from UCLA, researchers examined patients with FTD and Alzheimer's disease (AD). Typically, early in the illness FTD affects the front parts of the brain, while AD affects the back parts of the brain. When damaged, the front parts of the brain are more involved in poor judgment and a decreased control over one's actions. Researchers studied both groups for sociopathic behavior, evaluated the characteristics surrounding their acts, and compared the groups on neuropsychological tests and brain-imaging studies. There were twenty-eight patients in each group. Sixteen (57 percent) of the FTD patients had sociopathic behavior compared to two (7 percent) of the AD patients. Sociopathic acts among FTD patients included such things as unsolicited sexual acts, traffic violations, and physical assaults. When interviewed, the FTD patients with sociopathic acts were aware of their behavior and knew that it was wrong but could not prevent themselves from acting impulsively. They claimed remorse, but they did not act on it or show concern for the consequences. Among FTD patients with sociopathy, brain-imaging studies showed right prefrontal cortex involvement. The PFC helps us supervise our behavior and control the sneaky thoughts most of us have.

I have treated patients with FTD who started sexually abusing children. These men had no prior history of bad behavior. I have treated others with late-onset PFC disorders who developed unusual sexual behaviors. One case was particularly sad. A pastor of thirty years underwent brain surgery for a tumor to his PFC. He took time off from his position in the church. At first the operation seemed a success and the pastor went back to work. Then gradually over the next year, his behavior started to become bizarre. He had temper issues at church and was less reliable than before. Through his church he developed a friendship with a seven-year-old boy. Over time the relationship turned sexual. When he was caught, the whole community was stunned. The investigator for the case said there was absolutely no evidence of this type of behavior prior to the surgery. Yet, because of his position of trust, the judge gave this pastor thirty years in prison.

Tourette's Syndrome

Other areas of the brain besides the PFC can also be involved in sneaky behavior. The PFC helps us think about and supervise our odd behaviors. The basal ganglia and anterior cingulate gyrus can also fire abnormally and wreak havoc. Tourette's syndrome is an example that involves both of these systems. People who have Tourette's syndrome (TS) have uncontrollable urges to move their muscles (tics) or say exactly what is on their mind. They can control the urges for a while, but like tension on a rubber band that needs to be released, the urge builds until it has to be set free. TS is classified as a tic disorder where people have both motor (involuntary muscle movements) and vocal tics (involuntary vocalizations). Examples of motor tics include shoulder shrugging, leg movements, hip thrusts, excessive blinking, eyebrow raising, facial grimaces, head jerking, punching, and even sexual gestures. Examples of vocal tics include puffing, blowing, throat clearing, whistling, animal noises (barking, mooing, crowing), and swearing (termed *corprolalia*). Several years ago I spoke for the Tourette Syndrome Foundation of Canada on the island of Victoria. I spoke in front of four hundred people who had TS. It was more than interesting to speak in front of that many with tic disorders. In the audience there were people who were barking, whistling, and jerking. It taxed my ability to stay focused on my talk.

During the middle of the lecture someone blurted out, "Fuck you."

Taken aback, I just ignored him.

A few minutes later, it happened again: "Fuck you" came from the audience.

Now, I started to sweat. What should I do?

One more time, "Fuck you," came even louder.

I couldn't stand it anymore and said, "Is that a tic? Or don't you like the lecture?"

The man blushed and said it was a tic. But how is a speaker really to know?

TS is a treatable disorder, with medication and behavioral therapy. Without therapy it can ruin lives. David Comings, MD, a researcher at the City of Hope in Los Angeles, writes in his book *Tourette Syndrome and Human Behavior,* that families of Tourette's sufferers tend to have other unusual behaviors. These have included unusual sexual behavior, violence and abuse (especially within the family unit), obsessive compulsive tendencies, anxiety disorders, manic depression, and even psychotic symptoms. TS highlights that there are underlying mechanisms in the brain that correlate with impulse control disorders.

Obsessive-Compulsive Disorder

Obsessive-compulsive disorder (OCD) is similar to TS. In fact, about half the people with TS also have OCD. Obsessive-compulsive disorder affects somewhere between two to four million people in the United States. This disorder can impair a person's functioning and often affects a person's sexuality. OCD is often a disorder secretive to the outside world, but not to those who live with the person. The hallmarks of this disorder are obsessions (recurrent disgusting or frightening thoughts) or compulsions (behaviors that a person knows make no sense but feels compelled to do anyway). The obsessive thoughts are usually senseless, repugnant, and invasive; they are sneaky and may involve repetitive thoughts of violence (such as killing one's child), contamination (such as becoming infected by shaking hands), doubt (such as having hurt someone in a traffic accident, even though no such accident occurred), or sexuality (such as unusual acts with children or animals). Many efforts are made to suppress or resist these thoughts, but the more a person tries to control them, the more powerful they become.

The most common compulsions involve hand-washing, counting, checking, touching, and masturbating. These behaviors are often performed according to certain rules in a very strict or rigid manner. For example, a person with a counting compulsion may feel the need to count every crack on the pavement on their way to work or school. What would be a five-minute walk for most people could turn into a three- or four-hour trip for the person with obsessive-compulsive disorder. They have an urgent, insistent sense inside of "I have to do it." A part of the individual generally recognizes the senselessness of the behavior and doesn't get pleasure from carrying it out, although doing it often provides a release of tension.

The intensity of OCD varies widely. Some people have mild versions, where, for example, they have to have the house perfect before they go on vacation or they spend the vacation worrying about the condition of the house. The more serious forms can cause a person to be housebound for years. I once treated an eighty-three-year-old woman who had obsessive, sexual thoughts that made her feel dirty inside. It got to the point where she would lock all her doors, draw all the window shades, turn off the lights, take the phone off the hook, and sit in the middle of a dark room trying to catch the abhorrent sexual thoughts as they came into her mind. Her life became paralyzed by this behavior and she needed to be hospitalized.

New research has shown a biological pattern associated with OCD. Brain-SPECT studies have shown increased blood flow in the basal ganglia and anterior cingulate gyrus (ACG). The ACG is involved in allowing a person to shift his or her attention from subject to subject. When this area is overactive, a person gets "stuck" on the same thought or behavior.

Like most forms of psychiatric illness, OCD has a biological basis, and part of effective treatment often involves medication. At this writing there are eight "antiobsessive medications" and there are more on the way. Before 1987 there were no good medications to treat OCD. The current medications that have shown effectiveness with OCD are Anafranil (clomipramine), Prozac (fluoxetine), Zoloft (sertraline), Paxil (paroxetine), Effexor (venlafaxine), Luvox (fluvoxamine), Celexa (citalopram), and Lexapro (escitalopram oxalate). These medications have provided many patients with profound relief from OCD symptoms.

In addition, behavior therapy is often helpful for these patients. This is where a patient is gradually exposed to the situations most likely to bring out the rituals and habits. Behavior therapy also includes thought-stopping techniques and strong urging by the therapist for the patient to face his or her worst fear (for example, having a patient with a dirt or contamination fear play in the mud).

There is a group of disorders that have been labeled as Obsessive Compulsive Spectrum Disorders. It is based on the premise that these disorders occur because the person experiences repetitive unwanted thoughts or behaviors. They tend to get stuck on thoughts and cannot get them out of their minds unless they act in a specific manner. OCD spectrum disorders include trichotillomania (pulling out one's own hair), onychophagia (nail biting), Tourette's syndrome, kleptomania (compulsive stealing), body dysmorphic disorder (unreasonably feeling a part of the body is exces-

THE BRAIN IN LOVE · 123

sively ugly), compulsive shopping, pathological gambling, fetishes, and sexual addictions.

In the past decade the Internet has brought a whole new meaning to pathological gambling and sexual addictions. Both are on the rise with younger and younger people. The term *addiction* used to be used exclusive to chemicals such as alcohol, drugs, or nicotine. With recent research on the brain, we now understand that many behaviors can become as chemically addictive as a substance. Gambling and pornography can be such an addiction.

Pathological gambling includes all betting behaviors that interfere or hurt personal, family, or work-related activities. The essential features of a gambling addiction include:

- increased preoccupation with gambling
- a need to bet more money more frequently
- restlessness or irritability when attempting to stop
- "chasing" losses
- loss of control, manifested by continuation of the gambling behavior in spite of increasingly serious negative consequences
- in extreme cases, financial ruin, legal problems, loss of career and family, and even suicide.

According to the National Gambling Impact Study Commission, the national lifetime prevalence of gambling is no less than 1.2 percent of the total population (2.5 million). In longstanding gambling markets such as Nevada, more than 5 percent of the population will develop some problem with gambling, a prevalence rate about five times that of schizophrenia and more than twice that of cocaine addiction.

Youth are more troubled and addicted than adults. The prefrontal cortex (PFC) is not fully developed until age twenty-five, making it much more likely that teens and young adults will have problems controlling their impulses. According to the National Coalition Against Legalized Gambling, the following are the prevalence rates among youth groups: 16 to 24-year-old males, 4 percent; 11 to 18-year-old males, 4 to 7 percent; national average, all ages, 1.2 percent.

Sexual Addictions

Sexual addiction can encompass a wide variety of activities. Sometimes a sex addict has problems with just one undesirable behavior, sometimes with many. According to Sex Addicts Anonymous, a high percentage of sex addicts think that their unhealthy use of sex has been a progressive process. It may have started with excessive masturbation, the use of pornography, or a sexual relationship, but over the years it progressed to increasingly risky behaviors. The core problem with sexual addictions, like all addictions, is the feeling of powerlessness or helplessness over a compulsive behavior, resulting in a person's life becoming unmanageable or out of control. Addicts typically experience shame, guilt, and self-loathing. The addict tries to stop, but is unable to do so. The consequences are often severe, including terminated relationships, problems at work, arrests, financial troubles, a loss of sexual interest in anything not associated with the addiction, low self-esteem, and feelings of hopelessness.

The preoccupation with sex takes up tremendous amounts of energy, draining time from other activities and responsibilities; as this intensifies, behaviors or rituals follow, which usually leads to more troubled behavior, such as searching the net for pornography, lying about sex, or secretive flirting. There is usually guilt over the behaviors, followed by shame, despair, and confusion.

Here are six questions to consider if you think you may have a sexual addiction:

1. Do you keep dark secrets about your sexual activities from those who should know? Do you lead a double life?
2. Do you frequently feel remorse, shame, or guilt after a sexual encounter?

3. Is it taking more variety and frequency of sexual activities than previously to bring the same levels of excitement and relief?

4. Have you ever been arrested or are you in danger of being arrested because of your practices of voyeurism, exhibitionism, prostitution, sex with minors, indecent phone calls, etc.?

5. Do your sexual activities include the risk, threat, or reality of disease, pregnancy, coercion, or violence?

6. Has your sexual behavior ever left you feeling hopeless, alienated from others, or suicidal?

If the answer to any of these questions is yes, professional help may be in order. See an expert in sexual addiction.

In a study performed by my colleagues Mark Laaser and Richard Blankenship on sex addicts, the PFC was clearly involved. The researchers used the Amen Clinic Brain System Checklist, a questionnaire I developed based on our brain-imaging work, with seventy patients who met the diagnostic criteria for a sexual addiction. In addition, I performed brain-SPECT studies on eleven of the patients. The results of the survey were that 67 percent of the participants showed prefrontal cortex problems. Fifty percent of the participants had anterior cingulate gyrus issues (a tendency to get stuck on negative thoughts or behaviors, like addictions). There was also a high association between limbic (mood issues) and basal ganglia problems (anxiety issues). On the SPECT scans, eleven out of eleven patients with sexual addictions showed low activity in the PFC.

Here is an example of a sex addict. Joseph had been married eighteen years when he came to see me. During his marriage he had many affairs and spent excessive money in strip clubs and on Internet pornography sites. Even though he did not want to be divorced, and his wife was threatening to leave him if he didn't stop, he felt he had little control over his sexual behavior. Joseph had a family history of substance abuse and addiction, common among sex addicts. When he was arrested for soliciting sex from an undercover vice police officer, he sought treatment for his sexual addiction. Traditional psychotherapy was ineffective. Attention-deficit/hyperactivity disorder (ADHD), a neurological disorder that often affects the PFC and that is highly responsive to certain medical treatments, was suspected and he was sent to my clinic for an evaluation. His scans showed marked decreased prefrontal cortex activity. He was prescribed stimulant medication and placed on a brain healthy program (healthy diet, exercise, fish oil, and vitamins). With his medical treatment in place, he sought the help of an expert in sexual addiction. Joseph is now ten years sober from his sexual addiction. He and his wife have renewed their wedding vows and continue to do well.

Like Joseph, people with ADHD are likely to engage in some form of addiction. According to ADHD expert Wendy Richardson, "It is common for people with ADHD to turn to addictive substances such as alcohol, marijuana, heroin, prescription tranquilizers, pain medication, nicotine, caffeine, sugar, sex, cocaine, and street amphetamines in attempts to soothe their restless brains and bodies. Using substances to improve our abilities, help us feel better, or decrease and numb our feelings is called self-medicating." Treatment of ADHD and other underlying disorders affecting the PFC substantially increase the efficacy of treating addictive disorders.

Unusual Sex

In the movie *There's Something About Mary,* the main character, Mary, was stalked by Dom when she attended Princeton University. Dom's obsessive behaviors led to restraining orders. In addition, Mary had to change her name and buy all new shoes. Dom had taken all of her footwear. At the end of the movie Dom finds Mary once again. He is seen passionately clutching several pairs of her shoes as Mary's friends are pulling him off her. Dom is clearly over-attached to her shoes.

Can neuroscience help us understand these deviations, paraphilias, and fetishes? Unusual sex can have elements of both obsessive-compulsive disorder (repetitive thoughts or behaviors that feel as though they are out of control), and attention-deficit/hyperactivity disorder (associated with impulse control problems). Obviously, there is more to the story, as most people with OCD or ADHD do not engage in these behaviors. Imaging and brain chemistry findings can help uncover some the reasons behind these behaviors.

Sexual excitement is an individual experience; things that some people find stimulating are not the least bit interesting to others. Within a culture there are sexual norms that most individuals and couples incorporate into their own lives. Outside the range of these norms lies an area of more unusual sexual acts termed *paraphilias* or *perversions*: sneaky thoughts that get played out in actions. The word *paraphilia,* coined by Sigmund Freud, has Greek origins. "Para" meaning along the side and "philia" meaning love. It has been estimated that men outnumber women with a paraphilia 20:1, indicating that brain sex differences and hormonal factors are likely involved. The medical definition of a paraphilia connotes sexual deviance, perversion, or abnormality. It is seen as a medical problem only if the behavior interferes with relationships, health, work, or legal status or causes significant emotional distress. Culturally bound limits distinguish between normal and deviant sexual practices. Here is a list of common paraphilias and fetishes.

COMMON PARAPHILIAS

Exhibitionism: exposure of genitals to a stranger
Fetishism: use of nonliving objects, such as Mary's shoes
Frotteurism: rubbing against a stranger
Hyphephilia: fabrics
Klismaphilia: enema
Masochism: pain, humiliation, or punishment of self
Narratophilia: erotic talk
Pedophilia: children
Sadism: psychological or physical suffering of another

Stigmatophilia: body piercing or tattooing

Voyeurism: observation of others undressed or in the act of sexual activity

Transvestic fetishism: cross-dressing

OTHER TYPES OF PARAPHILIAS

Acrotomophilia: amputee partner

Asphyxiophilia: self-strangulation

Autonepiophilia: diaperism

Coprophilia: feces or defecating

Mysophilia: filth

Urophilia: urine or urinating

Zoophilia: animals

Fetishes

A fetish, defined as "an excessive or irrational devotion to some object or activity," such as Dom's attachment to Mary's shoes, becomes a problem when people become dependent on the object or practice. It can be an inanimate object, such as shoes, or an animate object, such as feet or breasts. People use various things to achieve sexual excitement. A fetish is when that object becomes the preferred or exclusive method of achieving excitement.

TYPES OF FETISHES

Shoes: boots, high heels

Clothes: leather, jackets or pants, suspenders, underwear, bras, raincoats, lace, silk, and nylon

Body parts: Buttocks, breasts, legs, elbows, ears, feet

Profiles in Fetishes

I Like to Watch

Watching unsuspecting people engaged in sexual activity or getting undressed holds an air of appeal to voyeurs because they are aroused by the chance of getting caught watching. They seek excitement, often as a way to stimulate an underactive brain. Exhibitionists find excitement in revealing their genitals to strangers, also considered an impulse-control disorder. A sixty-five-year-old man with dementia displayed this kind of behavior and was arrested five times, within one year, because of complaints regarding indecent exposure. Dementia often results in a disinhibition due to frontal-lobe deficits.

Vacuuming to Death

The tools of masturbation are as varied as people's imaginations. I have heard and read stories about vibrators, ice cream scoopers, butter knives, an old-fashion whisk, serving spoon, hairbrush, toothbrush, pillows, stuffed animals, pens, candles, dildos, tampons, towels, bedding, bottles, cell phones, curling irons, hot dogs, salami, Popsicles, and fruit. The list goes on and on. One case, however, stands above the rest in weirdness. A seventy-seven-year-old man was found dead in his bathroom due to a heart attack. Next to his body was found a vacuum cleaner and hair dryer. Both these electrical devices were still on and the man's underpants were actually found lodged in the hose of the vacuum! More than likely, these appliances were used autoerotically during masturbation.

Don't Try This at Home

People will go to almost any means to get sexual pleasure, even strangling or asphyxiating themselves to near death, or in some cases, death itself. For some, inducing a lack of oxygen to the brain while masturbating enhances sexual experience. The ways to choke off one's air supply are innumerable. For this purpose people have used hanging or choking devices, such as ropes, electrical cords, ties, belts; constricting devices around the chest or abdomen; plastic bags covering the face; toxic gases or chemicals that are inhaled; or partial or total submersion in water, known as aquaerotic asphyxiation. The problem with this sort of excitement is that hundreds of deaths each year result from this behavior. Sometimes the release mechanisms fail, sometimes people do not know when to quit. This paraphilia can affect people of all ages. In one study of 132 people, 6 were in their fifties, 2 in their sixties, and 1 in his seventies. This practice may continue throughout life; frequency may lessen with age, but the motivation behind the behavior may never be completely gone.

Stop Being a Baby

Paraphilic infantilism or autonepiophilia, also called diaperism, is the desire to be a baby and to talk and act the way an infant does. One thirty-five-year-old man employed in law enforcement was affected with this disorder his whole life. While at his job he did not feel like a baby and none of his colleagues knew about his desires, he dressed like a child when going out into public. His desire to become a baby began around the age of twelve. It can be postulated that the threat of sexual maturity caused regression. His infantlike behavior involved wearing diapers, eating baby food, sleeping in a custom-sized crib, and sucking on a pacifier. While wearing diapers he would urinate and defecate in them as a child would and, at other times, he would masturbate while wearing them. This case is an extreme side of a disorder that can be looked at as a spectrum ranging from infantile obsession to sadomasochistic behaviors involving domination and discipline.

Unusual Places

Like other orifices in the body, many objects have been used to stimulate the anus and rectum. The oldest reported case dates to the sixteenth century when a foreign object that was inserted into the rectum became lodged, and at that time removal of the object was not possible. To this day, physicians find themselves extracting objects from the rectum, sometimes requiring surgery. Such was the case of a sixty-three-year-old man who inserted a salami into his rectum. Twenty-six percent of those admitted to the hospital for foreign bodies in their rectums required surgery. Reports of patients seen in an emergency room for this problem vary and do not occur only in young people. In a large study of these patients seen at USC and UCLA Medical Centers between 1993 and 2002, the mean age was 40.5 years old. Other reports include an elderly woman at a nursing home who reportedly inserted utensils into her rectum. She said she was "digging herself out" and had been doing so for years, which caused her long-term constipation and stool incontinence. A sixty-five-year-old man who described abdominal pain had an inverted glass jar lodged in his rectum, which eventually required surgical removal.

Can You Say Mooo . . .

Zoophilia, or bestiality, is usually condemned as animal abuse and outlawed as a crime against nature. Scientific surveys, however, have shown that it is more frequent than most people would think. One to two percent, and perhaps as many as eight to ten percent, of sexually active adults have had significant sexual experience with an animal at some point in their lives. In the past, bestiality was particularly hated for the fear that it would produce monsters. Just over half of U.S. states explicitly outlaw sex with animals. In some countries laws existed against single males living with female animals. For example, an old Peruvian law prohibited single males from having a female llama. The Bible clearly prohibits bestiality. Leviticus 18:23 says, "And you shall not lie with any beast and defile yourself with it, neither shall any woman give herself to a beast to lie with it: it is a perversion." And 20:15–16 continues, "If a man lies with a beast, he shall be put to death; and you shall kill the beast. If a woman approaches any beast and lies with it, you shall kill the woman and the beast; they shall be put to death, their blood is upon them."

A commonly reported starting age for bestiality is at puberty, which is consistent for both males and females. Those who discover an interest at an older age often trace it back to memories or feelings that developed during this earlier period. As with human attraction, zoophiles may be attracted to only particular species, appearances, personalities, or individuals, and both these and other aspects of their feelings vary over time.

A colleague of mine from North Dakota once treated a man who had an obsession with heifers (young cows). It started when he was nine years old. While milking a cow, the scent and feel aroused him. Later that day he masturbated to the experience. Thirty years later he had had sex with more than five hundred heifers, but no human females. He had the cows perform oral sex on him by placing peanut butter or honey on his penis.

In October 2005, a Bangladeshi man was sentenced to three months in jail after pleading guilty to charges of bestiality in the United Arab Emerites. The camel involved in the case was put down in accordance with Islamic law. A court official said the man, who worked as a driver, had been spotted going into his employer's barn on a regular basis. His employer became suspicious as his duties did not involve him dealing with animals. The employer followed his driver into the barn one day and saw him starting to have sex with a female camel. The owner lost his temper and started beating him. He then took him to the police station to press charges. The driver confessed to police that he had fallen in love with the camel and had sex with the animal.

Anatomy of a Fetish and the Arousal Template

According to sex addictionologist Mark Laaser, PhD, "the arousal template" underlies many sexual fetishes and paraphilias. The theory of the arousal template says that it is important to understand where you were and how old you were when you experienced your first sexually arousing experience. This often lays the neural tracts for later excitement, even if the experience happened as early as age two or three. The first experience gets locked into the brain, and when you get older, you seek to repeat the experience because it was the way you had the initial arousal, like the first time you fell in love, used cocaine, or had to cope with pain. Here are several examples,

Adam, a forty-five-year-old stockbroker, grew up with an incredibly demeaning, shaming mother who gave him no love and acceptance. Four little girls lived next door to him when he was growing up. The little girls often came over to his house to play. They usually wore white socks and black patent leather shoes. When Adam got married, his fetish was that he wanted his wife to wear lacey dresses, patent leather shoes, and look like a little girl. It took him back to the arousal template of playing with the neighbors. His wife was initially okay with the sexual play, then not, as he wanted it every time to get an orgasm.

Gary also had a harsh and critical mother, plus an alcoholic dad. He grew up in Texas playing rodeo with a neighboring girl. They tied each other up as part of their play. It gave him a sense of control and pleasure, control he did not have over his erratically behaving mother or father. When he became sexually active as an adult, he wanted to tie women up, which eventually turned into an S&M fetish.

Lucy was an adult woman whose fetish was being spanked, slapped, and tied up. When she was five years old her father sexually molested her, including digital penetration. When she told her mother about the abuse, she was slapped because she didn't want to hear about it. The arousal template got mixed up with messages of pleasure and pain.

When Fred was a little boy, he was molested by an uncle. He felt ashamed and furious. As an adult he got married and loved his wife deeply, but also went to adult bookstores that had video booths. He would invite strange men into the booth and offer to do oral sex on them. When they dropped their pants, he would go into a rage and beat them up. His behavior was attempting to right a past wrong.

For children raised in the 1950s, there were some mothers who were overconcerned with their children's bowel regularity and gave them frequent enemas. If it was arousing to them, they would later insert objects into their rectum. I once treated a patient whose mother subjected him to frequent enemas. He was also a coffee addict. Twice a day he would brew coffee, put it in an enema bag, insert it into his rectum, and then masturbate.

At age thirteen, a boy was coming out of the shower naked and looked out the window toward next door. In the window he saw a woman washing the dinner dishes. She saw him and smiled at him. Subsequently, he would shower at the same time each night hoping to get a smile from the woman next door. As an adult he became a male exotic dancer and would also pose nude for art classes.

Chuck was a church pastor who got caught looking at porn, exclusively of Asian women. He had served in Vietnam in 1965 and 1966 as a medical corpsman. It was common for him to deal with death and disfigurement. For R&R he went to the massage parlors in Tokyo where he wanted only local women. He was emotionally locked into these women from dealing with the trauma and stress of wartime. Part of the arousal template occurs when you go through a stress, such as something that happened at an earlier point in life. Subsequent stress can trigger how the sexual arousal helped you escape the earlier turmoil.

The arousal template can happen later in life. For example, when a young college couple was driving home from a concert and the woman was masturbating her boyfriend with her feet; he ejaculated in front of a woman for the first time. Subsequently he became obsessed with feet. He collected woman's running shoes and took pictures of female marathons, but only of the women's shoes. He had a collection of two hundred pairs of female running shoes and wanted his wife to wear them as a part of sex. They came into treatment because she discovered him looking at Internet foot pornography sites.

Further Dynamics

One of the doctors in my clinic, Leonti Thompson, has extensive forensic experience. He was the former chief of psychiatry for the

Department of Corrections in the State of California. In his role with prisoners he examined and treated many sex offenders. Often his work involved trying to explain seemingly bizarre and/or offensive behavior in a way that could help to provide a basis for more objective dispositions of the inmates. He found that explaining these unusual behaviors from a developmental point of view was helpful. In this explanation he explained that libidinal/sexual drives—not yet given a definitive sexual categorization in the immature brain—could be attached to a variety of activities based on co-occurring emotionally laden experiences. These involved the tying together in the brain patterns that were out of the ordinary. An example comes from Victorian schools in which (probably sadistic) teachers would emphasize caning problem boys on their buttocks. As adults a number of these men would go to receive their "discipline" from the ladies in the leather boots and whips. He explained that the sexual receptive areas in the brain overlap the areas receiving stimuli from the rectal areas and with punishment an association was set up between those two areas. There was confusion then between pleasure and relief from pain. As the individual matured, the previously undifferentiated libidinal valences became more clearly defined as erotic. Pain and sex became conjoined in the mind and relief from pain and pleasure also became conjoined and confused. In this way one could understand how masochistic experiences might be related to sex.

Dr. Thompson has seen a number of patients at our clinics who have sexual problems involving legal considerations, and explanations to the families are helpful. He uses brain-SPECT imaging findings to explain relevant brain dynamics as the obsessive aspects represented by the anterior cingulate gyrus hyperactivity, for example. He recently had a patient who had a diaper fetish in a small town. Both he and his wife held visible positions and the public revelation was painful. The couple was severely traumatized. He had anterior cingulate behavior that was getting him into potentially serious consequences with the court. He was also very depressed. An important part of helping to bring a more therapeutic slant into the process was trying to have the couple see his behavior in a more logical way.

The Brain and Paraphilias

Three different areas of the brain are involved in paraphilias: prefrontal cortex (PFC), anterior cingulate gyrus (ACG) and basal ganglia (BG), and limbic or emotional structures. Each of these areas contributes in a different way and they may, in fact, act together in a circuit or system that reinforces the reoccurrence of the sexual behavior. Hormonal issues are important as well. Paraphilias are what I term an impulsive-compulsive disorder. They have features of both a lack of impulse control, that involves the PFC, and compulsion, that involves the ACG and BG.

The PFC, as we have seen, is critical to executive function. Damage to this area causes disinhibition and a lowering of impulse control. Therefore, people are more likely to behave in ways that are not acceptable and feel no qualms about their behavior. An article in the Archives of Neurology from University of Virginia researchers described a case of a man with a right prefrontal cortex tumor who developed pedophilia and was unable to inhibit his sexual urges despite knowing his behavior was wrong. The behavior resolved following tumor removal.

In a study from McLean Hospital in Boston, researchers evaluated 120 men with paraphilias, including 60 sex offenders. As a group, they were more likely to report a higher incidence of physical abuse (often associated with brain injuries), fewer years of completed education, a higher prevalence of learning and behavioral problems, more psychiatric/substance-abuse hospitalizations, increased work-related problems, as well as more lifetime contact with the criminal justice system. All of these issues are associated with low PFC function.

Overactivity in the ACG and BG are associated with compulsive behaviors and cause people to get stuck on negative thoughts and behaviors. In addition, a part of the BG called the nucleus accumbens is associated with pleasure sensations. When it is triggered by something perceived as pleasurable, it reinforces or encourages the behavior to happen again and again.

The limbic or emotional structures in the brain, including the hippocampus in the temporal lobe, add the spice or emotional charge to the paraphilia. The brain's memory centers work through association. One memory can trigger off a series of emotions, such as how hearing a song can trigger a happy memory of your lover, or the same song may trigger the sorrow of a failed relationship or deceased spouse. Likewise, certain paraphilias can trigger pleasure. There is a reported case of a sixty-five-year-old man who became obsessed with the desire for a peg leg. His fascination with peg legs began when he was young after seeing people with false legs. He thought they were fascinating (pleasure) and the prosthetics devices became more and more important over the years. It got to the point where he was aroused by them. He related the peg leg with positive feelings, generated from the brain's limbic system, so that whenever he would see or think about peg legs he would experience very positive feelings. He was unhappy at the time of his evaluation and believed that if he had a peg leg, he would be happy.

As we saw in Lesson Four, men and women have a number of important anatomical differences in the brain; one important area is called the hippocampus, part of the limbic system, on the inside aspects of the temporal lobes. Most paraphiliacs are men. In men, the hippocampus is known to be involved in the control of erection. There is evidence from animal studies that a larger hippocampus is related to polygamous behaviors. The hippocampus is involved in both mating strategies of these animals and their "geographic range." Paraphilacs are often diagnosed with antisocial personality disorder, have a larger number of sexual partners, and show greater geographic range than most men. Sociobiologists even distinguish their polygamous mating strategies than those of other people. Essentially, they are in a class of their own and the size of the hippocampus may have something to do with it.

Paraphilias have been reported with multiple sclerosis (MS). Multiple sclerosis is a disorder of the central nervous system where neurons lose their coverings and plaques become deposited into nerve-cell fields causing short circuits. MS affects multiple areas of the brain, including the prefrontal cortex. A wide variety of symptoms are associated with this disease, and usually it affects mood or cognition. There is a case report in the medical literature of a man recently diagnosed with MS who began displaying paraphilic behavior. He had no previous history of sexual abuse or dysfunction and was in a long-term relationship. As the MS worsened, he began to masturbate incessantly and sexually accosted strangers on the street. His behavior became so out of control that it eventually led to his incarceration. In his case, the changes in his brain precipitated the sexual behaviors. Paraphilic symptoms have also been associated with other brain diseases such as epilepsy, Parkinson's disease, dementia, and tumors.

Hormones play a significant role in sexual behavior. In the pursuit of treatments for paraphilias, research has shown that a common birth control method in women, Depo Provera, can help. This synthetic hormone decreases sexual desire and helps to tranquilize the brain. It has also been found to act in the temporal lobes and been used to treat epilepsy and rage attacks in men. It does not just lower plasma testosterone levels, implicated in hypersexual behavior; it also acts as a method of calming seizure like activity in the brain. In one case a thirty-eight-year-old man suffered brain damage after falling off his bike. He had no history of mental illness or sexual dysfunction and his wife thought he had been a good problem solver. After the accident, his behavior changed drastically and he began displaying hypersexual tendencies. He harassed his seventeen-year-old stepdaughter and she became the target of his sexual advances. His sexual behavior was related to his epileptic seizures and would occur simultaneous with them. After he received the hormone treatment, he stopped seizing and his sexual problems ceased as well.

Are Sex Offenders Treatable?

The treatment of sex offenders has changed radically in the last decade. What was once considered a hopeless disorder is now thought by some professionals to be treatable in many cases. The goal is to preserve normal sexual interests and behaviors while reducing deviant or paraphilic ones. According to Canadian psychiatrist John Bradford, pharmacological treatments have been shown to decrease the main problem in pedophilia, the preference of children for sexual gratification. Biological treatments, specifically castration and neurosurgery, have been used to treat sexual offenders to reduce their sexual drive and to prevent relapse. The studies of these treatments have reported markedly low recidivism rates, of about 5 percent following long periods of time. These outcome studies help us understand these disorders, which involve excessive response to male hormones, and what to do about it. The biological effects of surgical castration and male hormone suppression by antiandrogens and hormonal agents have the same effect on sexual behavior. In addition, Dr. Bradford has found antiobsessive medications, called SSRIs, which enhance serotonin, to be also helpful in this population.

Dwayne McCallum, a past medical director of a prison in Colorado, has used ideas from my brain-imaging research with sex offenders. He noticed that many of them had ADHD, and treating the disorder decreased their impulsivity and subsequently their recidivistic behavior. He also noted another group of sex offenders who had anterior cingulate gyrus issues (rigid, worried, inflexible, repetitive negative thoughts) that responded better to serotonin-enhancing medications, such as Lexapro. When it comes to sex offenders, most people think like my father, "You should kill all the bastards." My imaging work and the pioneering work of John Bradford, however, suggests a much more radical approach to them. By scanning and treating the abnormal brain function that many possess, we are likely to decrease subsequent crimes.

Getting Sneaky Thoughts out of Your Mind

As mentioned in the beginning of this chapter, we all get sneaky thoughts. Sometimes they turn into trouble. One way to rid yourself of the pesky or irritating thoughts is to play them out to their worst conclusion. Here is a case below from my *Men's Health* column. One man wrote in with the following question. "I walked in on my sister-in-law while she was changing. Now I can't stop thinking about her. What do I do?"

I wrote, "If you're thinking about what might happen the next time you're alone with your sister-in-law, let that fantasy play out in your mind. But when you do, give it a negative ending. Getting caught by your wife or sister-in-law's husband is generally a good start, but take the nightmare a step further by imagining the effect this would have on your relationships with other family members as well. It'll make the whole fantasy less appealing—and allow it to die a natural death. If the thoughts, or even paraphilias and fetishes, cause you distress, use this technique to decrease their frequency."

Lesson #7: We all get unusual or sneaky thoughts; that is normal. When they get out of control, it may be a sign that the brain needs help.

THE "OH GOD!" FACTOR

Sex As a Religious Experience

"If we accept the postulate given to me by Teresa during my Freshman year that, 'it will be a cold day in Hell before I sleep with you,' and take into account the fact that I slept with her last night, then I am sure that Hell . . . has already frozen over. The corollary of this theory is that since Hell has frozen over, it follows that it is not accepting any more souls and is therefore, extinct . . . leaving only Heaven thereby proving the existence of a divine being which explains why, last night, Teresa kept shouting 'Oh my God.' "

—FROM AN ANONYMOUS SOURCE ON THE INTERNET

God usually comes up at least once in the throes of sexual ecstasy. "Oh my God . . . Please God . . . God don't stop . . ." are not uncommon phrases preceding intense orgasms. On the surface it might seem as though religious ecstasy and sexual pleasure have little in common, besides, of course, calling out God's name. Looking under the surface, however, reveals some fascinating similarities. Experiences of God and sex are usually enhanced by rituals, music, and candles. Both experiences involve requesting help, if you consider "Oh God please don't stop" asking for help, and both experiences can be associated with tremendous joy.

In recent years imaging has been used as a tool for brain scientists to look at brain function in relation to religious experience and sexual ecstasy. Researchers have come to learn that behaviors that were seen as vastly different from each other have more in common than once thought. Both peak experiences seem to be processed primarily in the right side of the brain, especially the right temporal lobe and prefrontal cortex. So enhancing one experience may, in fact, help the other. Enhancing right-hemisphere function may enhance both religious and sexual experience.

Is there really evidence to support the connection between religious experience and sexual ecstasy? Sexual ecstasy or orgasmic pleasure appears to involve primarily the right hemisphere. Researchers from the University of Kuopio in Finland used brain-SPECT imaging to study eight healthy right-handed men during orgasm. They found overall decreased blood flow in the brain during orgasm except in right prefrontal cortex, where cerebral blood flow was increased significantly! There are twenty-three cases of temporal lobe epilepsy that are associated with the feeling of having an orgasm, called an orgasmic epileptic aura. Eighty-seven percent of the patients had their abnormality on the right side, especially the right temporal lobe. A study from the Department of Neuroscience at Norwegian University of Science and Technology in Trondheim, Norway, examined eleven patients with epilepsy who reported auras of ecstasy or pleasure. Four had erotic sensations and five described "a religious/spiritual experience."

Research from the Behavioral Neuroscience Laboratory at Laurentian University in Ontario, Canada, under the direction of Michael Persinger has studied religious experience and the brain for many years. They have reported that religious experience, especially of a sensed presence or being, can be induced by placing a weak magnetic field over the right temporal lobe. Women, who have greater access to the right hemisphere in general, reported more frequent experiences of a sensed presence than did men, and men were more likely than women to consider these experiences as "intrusions" from a negative alien source. Some of the phenomena generated by the right temporal lobe included "evil entities," gods, out-of-body experiences, and alterations in space-time.

Neurotheology

Rayford and Jill have both experienced strong religious visions. He was an agnostic and she was a Christian. He thought that the devil was following him; she thought the Blessed Mother appeared to her. Both have temporal lobe epilepsy (TLE). Like other forms of epilepsy, the condition causes convulsions but it is also associated with religious feelings, sexual ecstasy, and sometimes hallucinations. Research into why people like Rayford and Jill saw what they did has opened up a whole field of brain science: neurotheology.

Several religious figures are thought to have temporal lobe epilepsy. An example is Ellen White, one of the founders of the Seventh Day Adventist Movement born in 1827. She suffered a brain injury at age nine that totally changed her personality and she subsequently started to have powerful religious visions.

Author Fyodor Dostoyevsky was reported to have bouts of "temporal lobe seizures." He felt his affliction was a "holy experience." His biographer Rene Fueloep-Miller quotes Dostoyevsky as saying, that his epilepsy "rouses in me hitherto unsuspected emotions, gives me feelings of magnificence, abundance and eternity." In *The Idiot,* Dostoyevsky offers one of the most nuanced descriptions of the experience:

"There was always one instant just before the epileptic fit . . . when suddenly in the midst of sadness, spiritual darkness and oppression, his brain seemed momentarily to catch fire, and in an extraordinary rush, all his vital forces were at their highest tension. The sense of life, the consciousness of self, were multiplied almost ten times at these moments which lasted no longer than a flash of lightning. His mind and his heart were flooded with extraordinary light; all his uneasiness, all his doubts, all his anxieties were relieved at once; they were all resolved in a loft calm, full of serene, harmonious joy and hope, full of reason and ultimate meaning. But these moments, these flashes, were only a premonition of that final second (it was never more than a second) with which the fit began. That second was, of course, unendurable. Thinking of that moment later, when he was well again, he often said to himself that all these gleams and flashes of supreme sensation and consciousness of self, and therefore, also of the highest form of being, were nothing but disease, the violation of the normal state; and if so, it was not at all the highest form of being, but on the contrary must be reckoned the lowest. Yet he came at last to an extreme paradoxical conclusion. 'What if it is disease?' he decided at last. 'What does it matter that it is an abnormal intensity, if the result, if the sensation, remembered and analyzed afterwards in health, turns out to be the acme of harmony and beauty, and gives a feeling, unknown and undivined till then, of completeness, of proportion, of reconciliation, and of startled prayerful merging with the highest synthesis of life?' "

The brain feels joy and ecstasy. It also feels sadness and pain. Learning how to enhance brain function can enhance all areas of life, even the ones most sacred.

Sacred Sex

Sex can be a sacred act. Being inside another person's body, becoming one with him or her, allows for the exchange not only of bodily fluids, but also of energy forces, thoughts, and intentions. Sexual union can be a spiritual experience. Many religions of the world discuss sex in a sacred context.

Tantra (a Sanskrit word which means "woven together") is a term applied to several schools of Hindu yoga in which sex is worshipped. And tantra has been applied to sexual practices of other religions, including Tibetan Buddhism and Taoism.

Hindu Tantra

Tantra yoga is thought to date back thousands of years. There were rituals, or *pujas*, focusing on sex organs, such as *yoni puja*, a ceremony honoring the vulva, either of a statue or a living woman; and *linga puja*, honoring the penis, often in the form of a natural upright stone. Similar objects of worship have been found among the archaeological remains of many neolithic people around the world, leading some scholars to speculate that "sex worship" in some form or another is humanity's oldest religion. An Alaskan friend gave me a long, cylindrical carved statue. It was a walrus *ooskis* (penis). He said it symbolized fertility and was a traditional gift given to couples.

A major component of Hindu tantra is meditation stemming from a yoga tradition. In general, yoga is a Hindu practice that teaches practitioners to quiet activity of the body, mind, and will so that the individual may realize its distinction from them and attain liberation. Meditative yoga schools include hatha yoga (body posture) and bhakti yoga (devotion). Many yoga schools advocate a nonsexual approach to worship, in which visualizing a deity, chanting of a mantra, concentrating on symbols called yantras, and the practice of tapas (discipline) are the foremost activities. During a sexual approach to worship, yoni puja, the practitioner may meditate on a yantra—often a downward-pointing triangle—that symbolizes the vulva of the goddess.

Some tantra yoga teachers recommend meditative practices that also share elements with kundalini yoga, where subtle streams of energy are raised in the body by means of posture, breath control, and movements. Teachers in this school of tantra typically advocate retention of semen even during sexual excitement as a prerequisite for spiritual advancement.

Can meditation and prayer really enhance the brain and in turn enhance your sexual experiences, as suggested here? Research studies by Andrew Newburg from the University of Pennsylvania and others have shown that prayer and meditation can indeed influence brain function in a positive way. Studying Tibetan monks and Franciscan nuns, Dr. Newburg found that these meditative activities dampened the outside world (parietal lobes) and brought the focus internally by enhancing the prefrontal cortex.

With the sponsorship of the Alzheimer's Prevention Research Foundation, my colleagues and I teamed with Drs. Dharma Singh Khalsa and Nisha Money to study the impact of meditation on the brain. We chose a simple twelve-minute form of kundalini yoga meditation, Kriya Kirtan, that is easy for busy people to practice. It is based on the five primal sounds: *saa, taa, naa, maa* (*aa* being the fifth sound). Meditators chant each sound as they consecutively touch their thumb to fingers two, three, four and five. The sounds and fingering are repeated for two minutes out loud, two minutes whispering, four minutes silently, two minutes whispering, and two minutes out loud. We performed SPECT scans at rest one day and then after meditation the next day. We saw marked decreases in the left parietal lobes (decreasing awareness of time and space) and significant increases in the prefrontal cortex (which showed that meditation helped to tune people in, not out). We also observed increased activity in the right temporal lobe, an area that has been associated with spirituality and sexuality, which our meditators found amusing. "Of course," one said, "that is why we meditate."

Best known in the West are the several forms of tantra yoga in which worship services take the form of sexual rituals featuring slow, nonorgasmic intercourse as a prelude to an experience of the divine. This broad category of tantric sex ritualism, which derives from the pre-Hindu religions of Shaktism and Shaivism, has in turn produced two schools of practice: the "right hand path," a meditational, monogamous rite; and the "left hand path," in which dozens—or more—couples engage in the ritual sex act at the same time, sometimes following the lead of a pair of teachers.

Tibetan Buddhism, Taoism, and Tantra

A version of tantrism can be found in contemporary Tibetan Buddhism, where a blend of pre-Buddhist goddess worship is interspersed with rituals from the ancient Tibetan religion Bon. Like Hindu tantra, Tibetan Buddhist tantra encompasses schools of practice that range from the meditational to the sexually active.

Taosim has several tantric schools with different views on the role of sexual activity. One form of Taoist tantra, sexual alchemy, places emphasis on the search for a long life. Taoist tantric alchemy involves breath and muscle control and emphasizes the retention of sperm as proof of spiritual attainment. Other Taoist tantra teachers, working out of a paradigm that seems to be derived from Shaktism, claim that Lao Tzu, the founder of Taoism, was in fact advocating a form of yoni puja or worship of the vulva when he wrote about "the valley spirit."

Religious and Sexual Ecstasy: Practical Applications

If the right side of the brain is involved in both religious and sexual ecstasy, how can you use this knowledge to enhance your own and your partner's sexual experiences? Since the right side of the brain helps us process music and rhythm, soft tunes and a romantic dance may help you and your partner get in the mood. The right side of the brain also sees in pictures, instead of in words, so being in a beautiful environment is usually another plus. The right side of the brain also tends to be the anxious or nervous side of the brain, so calming worries, through a warm bath, relaxing back or foot rub, or a reassuring talk, can be helpful.

You can also stimulate the right side of the brain by kissing the left side of the body. This intervention can be tricky, because sometimes the right side of the brain is associated with anxiety and it might be better to kiss the right side of the body to stimulate a person's left side. I recommend you run a series of experiments on your partner's body to see what side he or she likes stroked or kissed better. Actually, this principle of experimenting is one of the best strategies you can use. See what turns your partner on . . . write it down . . . talk about it . . . and remember it.

Creating Rituals: Practical Applications

Creating sacred rituals for your sex life can yield many long-term advantages. These rituals do not need to be rigid, which stifles creativity, but rather can provide a fun, safe, intimate environment in which to more fully explore your sexuality. Start by setting the boundaries of what is acceptable behavior; talk about what is comfortable and not. It is important for you and your partner to discuss the things that make you feel relaxed, comfortable, excited, and sensual. It is important to create a safe environment in which to play. Some partners do this by agreeing on a code word which tells the other that they want to stop or are feeling uncomfortable, without breaking the mood. As well, it's important for you and your partner to talk about the things that make you feel good, uncomfortable, or that you are just plain unsure about. Sex, at its core, is a form of communication between two people.

After you have agreed to safe boundaries, you can take sex to a new level by investing in a few books or magazines. I write for *Men's Health* magazine and it is always filled with great sex tips for couples. *Cosmopolitan* and other magazines have playful ideas as well. Books on tantric sex or role-playing games can also be fun. Page through them together, talking about and exploring new ideas and techniques for inspiring passion.

Once you've agreed on the existing and new areas of interest to play in, begin creating your plan of action for that evening (or afternoon or morning). Often, simple rituals can be very exciting, such as a beautiful bath filled with rose petals and floating candles, along with sensual music, titillating aromas, oils for massage, and an aphrodisiac tea or treat. This alone can create a mood of irresistible desire. There are times, though, that more excitement is required to bring on the feelings of arousal. This is where doing something outside of the norm can take a really exhilarating form. For instance, going to the ocean or a lake in a private spot and skinny-dipping, pulling the stop switch on an elevator (that doesn't have surveillance cameras) and making love in a unique setting, or playing out a sexual fantasy. This is an area of unlimited possibility and fun, if both partners can agree to cooperate with each other on hidden desires. Again, the word *cooperation* is a very important one. It's crucial for both partners to feel comfortable but also be willing to be a bit flexible, as long as the method doesn't compromise a core value or involve unwanted pain for a partner. Sometimes by opening oneself up to new possibilities, whole new worlds of sensuality are unleashed.

Remember, ritual is symbolic of the experience to come, so it can take just about any form—a phone call to your partner at work, homemade cookies as an afternoon snack, or a walk through the park holding hands. This is your unique bonding which takes place in order to open up the channels for deeper love and connection between you and your significant other. Let go, be creative, and use your brain to have fun.

Lesson #8: Sex can be a spiritual experience.

BRAIN SEX TRICKS

Sex Potions, Passion, and
Finding the "Damn Spot"

"Just find the damn spot. All you have to do is find the spot."
—JOY BEHAR, *The View*

A ccording to Greek myth, Uranus, the father of the Titans, was castrated by his son Cronus. When his severed genitalia was thrown into the sea, the waters began to churn. From the sea foam (known in Greek as *aphros*), Aphrodite was born, the goddess of love and the mother of Eros (known by his Roman name as Cupid). This is the origin of the name *aphrodisiacs,* the term that describes the many methods reported to enhance libido, potency, and sexual pleasure. This chapter explores the neuroscience of the foods, drinks, drugs, scents, or devices believed to enhance sexual interest and performance.

Aphrodisiacs

Viagra, and his pharmaceutical brothers, Cialis and Levitra, have drawn international public attention to aphrodisiacs. You cannot turn on the television or listen to a radio show without hearing about these medications. We are constantly reminded of our desire for sex. Images of baby boomer men grace the screen as they relax by the beach. The suggestive tag lines include "Let the dance begin" and "Relax and take your time." Viagra even has a frequent-buyer card and you get your seventh prescription for free. It may seem as though there is a new obsession with sex, but mankind has been focused on sex—and ways to make it better— since the beginning of time.

The search for aphrodisiacs dates back at least five thousand years and has included useful remedies as well as ones that have hurt and killed people, and cost many animals their lives. For example, there are two reports on the effects of swallowing a substance containing toad secretions. Of seven previously healthy men who took the drug, the active ingredient in the West Indian "love stone" and the Chinese medication "chan su," four died. Toad secretions contain a substance that causes abnormal heart rhythms in people, which may be responsible for the high they feel when taking it. Unfortunately, it can also stop the heart fatally. Which aphrodisiacs are helpful, practical, and have research to show efficacy, and which are dangerous or illegal?

One aphrodisiac to avoid is Spanish fly (cantharidin), which has been used for a thousand years as a sexual stimulant. It is one of the best known and infamous modern love potions. Made from the dried body of blister beetles, so named for causing blistering of the skin, it irritates the urinary tract, sending a rush of blood to the genitals, causing the feeling of excitement. However, it is also a poison and illegal in the United States. While most commonly available preparations of Spanish fly contain cantharidin in negligible amounts, if at all, the chemical is available illicitly in concentrations capable of causing severe toxicity. Symptoms of cantharidin poisoning include burning of the mouth, trouble swallowing, nausea, blood in the urine, and painful urination. Priapism (painful erections that won't go away), seizures, and heart abnormalities are also possible. The practice of eating live beetles in Southeast Asia and "the kissing bugs" (triatomids) in Mexico may work in a similar way. Similarly, these bugs are also dangerous and should not be ingested.

For increasing libido, ambrein, a major constituent of a gray, waxy secretion (ambergris) found in the digestive tract of some sperm whales, is used in Arab countries, but it is illegal in the United States. This substance increases the concentration of several hormones, including testosterone. It was once used as a fixative in perfumes, and is used as an aphrodisiac in some parts of the world. When given a dose of ambrien, male rats got a lot more interested in sex, even when no female rats were nearby. When receptive females were put into the cages, the male rats had "recurrent episodes of penile erection, a dose-dependent, vigorous, and repetitive increase in intromissions and an increased anogenital investigatory behavior." The rats went into a mating frenzy. The researchers from King Saud University in Riyadh, Saudi Arabia, concluded that ambrien is indeed a sexual stimulant, adding legitimacy to similar claims by Chinese traditional medicine. Historically, Chinese noblemen would drink ambrien dissolved into a draught. Ambrein is illegal in the United States, where sperm whales are listed as an endangered species.

Asian ginseng is commonly used to treat sexual dysfunction in men and is available in the United States. Recent lab studies in animals have shown that both Asian and American forms of ginseng enhance libido and sexual performance. The effects of ginseng have several probable causes. There is a direct effect of ginseng on the central nervous system and genital tissues. There is good evidence that ginseng can facilitate erection by directly increasing blood flow to the penis. The effects of ginseng appear to work through the release of nitric oxide (similar to Viagra). Ginseng also affects the brain and has been shown to enhance the activity of neurotransmitters and hormones involved in sexual behavior. Ginseng is considered an adaptogen that enhances physical performance, promotes vitality, and increases resistance to stress and aging. When used appropriately, ginseng appears to be safe. Nevertheless, documented side effects include hypertension, diarrhea, restlessness, and vaginal bleeding. Most published research studies have used a standardized Panax ginseng extract in a dosage of 200 mg per day. Other sources recommend 0.5 to 2 g of dry root per day on a short-term basis, with the ginseng taken in tea form or chewed. Capsule formulas are generally given in a dosage of 100 to 600 mg per day, usually in divided doses.

The prettiest brains I have seen are those on ginkgo. Ginkgo biloba, from the Chinese ginkgo tree, is a powerful antioxidant that is best known for its ability to enhance circulation, memory, and concentration. The best studied form of ginkgo biloba is a special extract called EGB 761, which has been studied in blood vessel disease, clotting disorders, depression, and Alzheimer's disease. There are also reports that it enhances sexual function. Many psychiatrists use ginkgo to counteract the sexual side effects of SSRI antidepressant medications. There are many different forms of ginkgo, making dosing confusing. In the United States Ginkoba and Ginkgold (Nature's Way) are brands that have been compounded to reflect EGB 761. The usual effective dose is 60 to 120 mg twice per day. There is a small risk of bleeding in the body, and the dosages of other blood-thinning agents being taken may sometimes need to be reduced. Before taking supplements, make sure to talk with your doctor.

Yohimbine, extracted from Yohimbine bark, can facilitate erections by stimulating the nervous system and increasing blood flow to the penis. But it is not for everyone. Common side effects include increased heart rate, raised blood pressure, anxiety, and nausea. Available by prescription, Yohimbine should be used under a doctor's supervision.

L-arginine is a naturally occurring amino acid that is a precursor to nitric oxide. It has been used with some success to enhance sexual function because it is believed to improve blood flow to the genitals. Researchers from the University of Copenhagen in Denmark reported that in rats L-arginine helped regulate vaginal smooth-muscle tone and also affected blood flow in all areas. L-arginine coupled with yohimbe has been found to make a positive difference in women with low sexual desire. Researchers from the University of Texas, Austin studied twenty-four women using this combination and found it significantly increased blood flow an hour after they were given these substances.

Sexy Scents

Humorist Dave Barry once wrote, "Of all the human senses— sight, hearing, touch, taste, and the feeling that a huge man with a barbecue fork is lurking in the closet—perhaps the least appreci- ated, yet most important, is our sense of smell." In treating patients who suffered from loss of the sense of smell, psychiatrist Alan Hirsch, from the Smell and Taste Treatment and Research Foundation in Chicago, found that almost 25 percent had also developed sexual dysfunction, suggesting that odor can have a large impact on sexual arousal. Your sense of smell is the strongest of your five senses and highly involved in sexual function, pleasure, and irritation. Scientists discuss the concept of "smell print," where memories are associated with certain smells. Years later, a smell will vividly cause a person to recall the memory associated with it. Smells are like fingerprints, highly individualized.

The deep limbic brain is involved with our sense of smell and interest in sex. The two go hand in hand. The smell of cooked cin- namon, on the one hand, is a natural aphrodisiac for men. The sex organ in the brain that is responsive to sexual-interest hormones is two and a half times larger in men than in women. When I told my mother about cooked cinnamon, she hit her forehead and said, "That is why I have seven children! He would never leave me alone." The Lebanese cook with a lot of cinnamon.

Dr. Hirsch has studied many smells associated with sexual interest in both males and females. Measuring penile blood flow, with a device that looked like a small blood pressure cuff, he found that certain scents or combinations of scents activated the erector machinery more than others. The combination scent of lavender and pumpkin pie was at the top of the list, increasing penile blood flow by 40 percent! Other winning scents included doughnut and

THE BRAIN IN LOVE · 155

black licorice, doughnut and pumpkin pie, doughnut and lavender, orange, cheese pizza, roast beef, and cinnamon buns.

Which ones work for females? To uncover this sweet-smelling piece of information, Dr. Hirsch measured vaginal blood flow with a special monitoring device. Increased vaginal blood flow is a sign of sexual arousal in women, while decreased blood flow is the opposite. He started by measuring the response to men's colognes. All of them decreased vaginal blood flow. Don't waste your money.

In my lectures, I ask the audience what scents they think most increase vaginal blood flow and sexual interest. Audience members blurt out some interesting answers, such as sweat, chocolate, vanilla, coffee, strawberry, and freshly minted money. The most interesting audience response came from a stunning, redheaded woman in Atlanta, who, in a very sweet southern accent, said, "The smell of fresh leather from a new Mercedes Benz." Very few crowds get the answer right. One of the best scents according to Dr. Hirsch's research was baby powder. "Baby powder!" I hear people in the audience say in amazement. "Why baby powder?"

The brain works through association, which is why certain smells or sounds trigger powerful feelings or memories, such as the smell of fresh baked bread taking you back to the comfort of Grandma's kitchen when you were a child. What do women associate with baby powder? Freshly diapered, cute little babies. Then, unconsciously, they want one; thus, the increased vaginal blood flow and interest! Dr. Hirsch also found that cucumber, licorice, lavender, and pumpkin pie all increased vaginal blood flow, while the smell of cherries and barbecues had the reverse effect.

Despite Dr. Hirsch's research, many ancient societies believed perfumes were aphrodisiacs and some new research suggests they may have been right. The Romans and Egyptians used lavish amounts of perfume containing musk. The musk used was from the anal glands of the Ethiopian civet cat. Current research suggests that the scent of musk closely resembles that of testosterone, the hormone that enhances a healthy libido in both men and women.

In scent studies performed at Toho University in Japan,

Professor Shizuo Torii showed the impact of floral and herbal essential oils on the nervous system. Sexual arousal and response is controlled by the two parts of the nervous system: the sympathetic nervous system (SNS), which prepares us for physical action or emergencies, also called the fight-or-flight system; and the parasympathetic nervous system (PNS), which stimulates relaxation. For those who need to relax in order to get in the mood of sex, the PNS should be dominant, while those who need to be stimulated would do better by enhancing the SNS. Professor Torii found that the PNS was stimulated by the scent of sandalwood, marjoram, lemon, chamomile, and bergamot. The SNS was increased by the scents of jasmine, ylang-ylang, rose, patchouli, peppermint, clove, and bois de rose. Aphrodisiacs need to be tailored to individuals, not large groups. Find the scents most appealing to you. Buy the essential oils of these substances in health food stores and dilute them by putting a few drops into a carrier like olive or canola oil.

Sexy Foods

A romantic meal can put you in the mood for love. The intimacy created by candlelight and soft music is important, but food also plays a starring role in our sexuality. The sweet texture of melted chocolate, the sweet juice and fragrance of an orange, the shape and smell of a cucumber—all stimulate the taste buds and the imagination. Additionally, several societies regard foods like bananas, cucumbers, asparagus, and carrots as erotic stimulants because of their penislike shape. The ancient Aztec name for avocado was *ahuacatl,* or testicle, because of the fruit's appearance. Virgin girls were forbidden from going outdoors during harvests of avocados. Today, researchers have verified that some of the ancient aphrodisiac foods do in fact contain certain vitamins and minerals that contribute to a healthy reproductive system and perhaps even a healthy libido. Here is a list of several potentially sexy foods.

Almonds (or nuts in general)

Almonds are a major source of essential fatty acids. These are vital as they provide the raw material for a man's healthy production of hormones and help the brain work better. The smell of almonds has also been reported to arouse passion in females. Almonds enhance phenylethylamine (PEA) production, as does chocolate, to increase brain stem activity, enhancing alertness. Try lighting some almond-scented candles to encourage your partner's mood and snacking on some (but not too many) yourself to store up energy for your performance. Eat them raw (with no added salt or sweetness), or, crush some fresh almonds and sprinkle them on your salad to get the energy you need.

Apples

Apples have been given as gifts of reverence for years, be it to teachers, friends, or loved ones. They actually belong to the Rose family and have been used in Yuletide rituals to symbolize fertility in nature. Apples not only clean the teeth and inspire the flow of saliva, but they also sweeten the breath, which is always a wonderful stimulus for the foreplay of kissing.

Artichokes

Artichokes contain a substance by the name of cynarin, which helps to strengthen the liver. Artichokes were traditionally used by the French as aphrodisiacs.

Asparagus

Many foods thought to be aphrodisiacs were considered so because of their phallic shape. Asparagus, however, has more than suggestive form. It is rich in vitamin E, a vitamin considered to stimulate production of our sex hormones and that may be essential for a healthy sex life.

Avocado

The Aztecs called the avocado tree *ahuacatl* or "testicle tree." While avocados can indeed look like that body part, they contain high levels of folic acid, which helps metabolize proteins, thus giving you more energy. They also contain vitamin B_6 (a nutrient that increases male hormone production) and potassium (which helps regulate a woman's thyroid gland), two elements that help enhance both male and female libido. In addition, avocados are loaded with phenylethylamines, even more so than chocolate. Plus, they have omega-3 fatty acids that help the brain work better, so you'll be more likely to get lucky.

Bananas

Bananas contain the bromelain enzyme, which is believed to improve male libido. Additionally, they are good sources of potassium and B vitamins such as riboflavin, which increase the body's overall energy levels.

Cabbage

Cabbage is wonderful for helping to increase circulation, therefore stimulating sexual energy.

Celery

While it may not be the first food that comes to mind when it comes to sex, celery can be a fantastic source food for sexual stimulation, as it contains androsterone, an odorless hormone released through male perspiration that turns women on.

Chili Peppers

Chilies may heat up your sex life, too, due to capsaicin—the substance that gives kick to peppers, curries, and other spicy foods. Capsaicin stimulates nerve endings to release chemicals, raising the heart rate, making us sweat, and possibly triggering the release of endorphins, giving you the pleasurable feeling of a natural high that is conducive to love-making.

Chocolate and Cheese

In addition to the silky texture of melted chocolate, this quintessential lovers' gift contains PEA, considered by some researchers as the "love chemical"—it imparts a feeling of well-being and excitement, like endorphins giving a natural high—and theobromine, a substance very similar to caffeine. Do not give too much chocolate, however. A little helps your partner focus, a lot gives him or her low blood sugar, which may put them to sleep. Cheese actually contains more PEA than chocolate. So order a cheese platter after dinner and see if it gets you both in the mood. If you still prefer chocolate, make it dark rather than milk. It has a much higher cocoa-solids content (and therefore more feel-good chemicals). To get him or her ready for sex in the morning, make your partner cereal with chopped apple and almonds—both have high levels of PEA, which will help get him or her in the mood.

Damiana

Damiana, or wild yam, has a traditional use as an aphrodisiac and chemical analysis shows that it contains chemicals that can increase sensitivity in the genitals. Damiana also has a reputation for inducing erotic dreams, when drunk at bedtime, although it has not been proven in clinical trials.

Eggs

Although not the most sensual of foods, eggs are high in vitamins B_5 and B_6. These help balance hormone levels and fight stress, two things that are crucial to a healthy libido. Eggs are also a symbol of fertility and rebirth. Some people will eat raw chicken eggs just prior to sex to increase their libido and maximize energy levels;

however, please be aware of the possibility of salmonella contamination in uncooked egg products. All bird and fish eggs contain B_5 and B_6.

Figs

High in amino acids, figs are believed to increase libido. They can also improve sexual stamina. Furthermore, the shape of a fresh fig and its sweet, juicy taste are two tangible aspects that are highly pleasurable to the human senses.

Garlic

Yes, you might need to stock up on some extra breath mints, but it'll be worth it. Garlic contains allicin, an ingredient that increases blood flow to the sexual organs. As such, it's a highly effective herb for increasing libido. If the odor just won't work for you, or you can't stand garlic, you can always take garlic capsules instead.

Nutmeg

Nutmeg has been mentioned in Indian Unani medicine for enhancing desire. When researchers from Aligarh Muslim University in India gave an extract of nutmeg to different groups of male rats daily for seven days, the female rats involved in the study were made receptive by hormonal treatment. The general mating behavior, libido, and potency were studied and compared with the results produced by Viagra, showing a significant augmentation of sexual activity. It increased erections and mounting frequency. Nutmeg was without any adverse effects.

Oysters

Oysters have long been considered the food of love, and legend has it that Casanova ate dozens of oysters a day, once even seducing a virgin by sliding an oyster from his lips to hers. Whatever the case, the truth is that oysters are high in zinc, a mineral used in the production of testosterone. Not only the hormone behind the male sex drive, testosterone is believed to stimulate the female libido as well. Oysters also contain dopamine, a chemical that increases focus and motivation for fun.

Sea Vegetables

Sea vegetables such as dulse, kelp, and nori are great aphrodisiacs because they are chock-full of minerals such as calcium, iodine, and iron, which help in balancing the thyroid gland and endocrine system . . . as a result, strengthening the libido.

Semen Taste Tips

Cinnamon, cardamom, peppermint, and lemon will improve the taste of semen. Garlic, onions, curry, or asparagus will have the opposite effect. Since women have a more sensitive sense of smell and taste, great hygiene is usually the best policy to reinforce the experience positively.

Delicious Vulvas

Some men say they do not want to give oral sex to women because of the taste or smell. The variances of tastes and smells depend on a number of factors, including what a person has eaten, where women are in their menstrual cycles, and the pH of the vaginal secretions. Of course, what a person finds appealing is very individual. Natalie Angier writes, in *Women: An Intimate Geography*, that our skin's pH stays between 6.0 and 7.0, while a healthy vagina is between an acidic 3.8 and 4.5. Examples of acidic pH foods are lemons (2.0), coffee (5.0), and wine (4.0). As a woman's pH climbs, her secretions smell stronger and stronger. Bathing with a hypoallergenic soap that has a low pH is one way to counteract the strong scent. Strong foods such as asparagus and garlic will increase the scent, as will multiple vitamins. Lemons, oranges, and grapefruits may soften and sweeten the scent. As with everything else about sex, tactfully talk about what you like and what you don't. Showering together before oral sex is often an erotic form of foreplay.

Exercise

Exercise can keep your heart healthy, your body slim, and your psyche sound, and now studies show it can act as an aphrodisiac, too. Although you may not feel so sexy after a sweaty workout, don't be surprised if you find yourself feeling in the mood for love. Research now suggests that along with all of the other health benefits exercise imparts, it can also give a big boost to your sex life. The reason has less to do with getting stronger than with the release of endorphins in the brain (as a result of physical exertion) that influence how we feel. These are the same neurochemicals responsible for a "runner's high" or the sense of exhilaration that comes from skiing down a mountain or after an intense aerobics class. It turns out these brain chemicals may also be linked to the release of hormones that power the sex drive. Research has shown that women who exercise regularly tend to have more active sex lives, are more easily aroused, and reach orgasm more quickly than those who don't work out. Less vigorous Eastern forms of exercise take another approach. Instead of stimulating brain chemicals to rev up the sex drive, yoga and tantra provide postures designed to help resolve specific sexual problems. Some are said to work by stimulating blood flow to the genital area while others are directed toward maximizing sexual performance and satisfaction. While there's no scientific proof that these ancient exercises add up to effective aphrodisiacs, practitioners seem pleased with the results. If regular exercise has no appeal, then twirling around on the dance floor can get the juices flowing just the same. In fact, whether the music is a minuet or mambo, moving to the beat is an age-old, socially sanctioned mating ritual. Whatever your preference—from the treadmill to the dance floor—it seems clear that the more you move your body, the better your sex life can be.

Touching, Feet, Foreplay, and a Very Strange-Looking Little Man in the Brain

Touch is a central aspect of sexual experience. Touching stimulates oxytocin, which in turn stimulates trust and comfort, setting the mood for sex. Knowing where to touch can be tricky. Not all areas of the body are created equal when it comes to sensory neurons, and starting with the genitals is usually considered bad form, even though they have a bazillion nerve endings.

Neuroscience can give us clear guidelines on how and where to touch someone to maximize erotic stimulation. Our skin has receptors that detect whenever something touches us. These "touch" receptors feed into the brain's parietal lobes in such a way as to create a "touch" or sensory map. Certain areas of the brain have many more touch receptors than others. In medical school I was introduced to the concept of the homunculus, a very strange-looking little man that illustrated the percentage of sensory nerve fibers in the brain (see Figure 9.1). From the illustration, you can see that the lips, hands, feet, and genitals get the lion share of brain space for sensation. The fingertips, for example, have the highest density of receptors: about 2,500 per square centimeter! Using this information, one can see why holding hands, kissing lips, and stroking a face with fingertips can be very stimulating. The brain has wired these areas to be very sensitive. In addition to the size of a sensory area, its placement and next-door neighbors can also give us important clues. For example, in the homunculus, the genital area is next door to the foot-sensation area. These areas share neural crosstalk and exchange information. This may be why many women say, "If you want your way, rub my feet." This anatomy lesson may also help explain foot fetishes, why women collect shoes, and why rubbing and kissing feet is so erotic.

"Feet have always had an erotic connotation," says Suzanne Baldaia, a fashion historian and professor from Rhode Island in an interview with the *Chicago Sun-Times*. "Folklore offers us many examples of the shoe being a symbol of female genitalia. In many cultures, a bride's shoe was offered to the bridegroom as a symbol

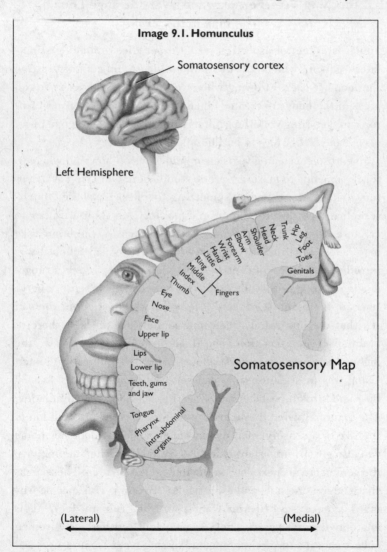

Image 9.1. Homunculus

Somatosensory cortex

Left Hemisphere

Somatosensory Map

Neck
Head
Shoulder
Arm
Elbow
Forearm
Wrist
Hand
Little
Ring
Middle
Index
Thumb
Eye
Nose
Face
Upper lip
Lips
Lower lip
Teeth, gums and jaw
Tongue
Pharynx
Intra-abdominal organs
Fingers
Trunk
Hip
Leg
Foot
Toes
Genitals

(Lateral) ← → (Medial)

of property, and as an exchange of something a little bit deeper." The prince in the story *Cinderella* is a classic example of a man chasing after a woman's feet. Feet are definitely an erogenous zone. Men love spiked heels, patent leather shoes, a glimpse of toes, which some have labeled toe cleavage.

When treating sexual dysfunction in women, Masters and Johnson found that sensual touching helped to focus on the pleasure associated with touch rather than genital response. They suggested foot rubs because they are not threatening and more comforting. Little did they know about the real reasons foot rubs worked. Foot rubs have also been helpful in pain management. Massaging feet stimulates the receptors in the muscle that activate the "nonpainful" nerve fibers and therefore prevent pain transmission from being felt. In studying this phenomena, researchers found that a twenty-minute foot massage after taking pain medication significantly decreased pain intensity and distress as compared to receiving medication alone.

Just Find the Damn Spot

During the middle of my appearance on *The View*, discussed in Lesson Four, where we were talking about the differences between male and female brains, Joy Behar interrupted me, saying, "Just find the damn spot. All you have to do is find the spot." She then repeated her observation several more times. I suspected she was referring to the clitoris, or perhaps even the G-spot. So, is it true? Do men just need to find the spot and rub up against it? I suspect not; sexuality is more complicated. But finding the spot is an important part of the puzzle, for both men and women.

Where is the G-spot? Is it real? The concept of the G-spot (more accurately referred to as an area, instead of a spot) was first discussed by Ernest Grafenberg, MD, in 1944. It was expanded on and popularized by Beverly Whipple and John Perry in the 1980s. Practitioners of tantric sex have been talking about this "sacred spot" for more than a thousand years. Even though it remains controversial, many women swear by it. To be clear, it is important to note that Dr. Grafenberg did not think there was just one erogenous spot. He wrote, "Innumerable erotogenic spots are distributed all over the body, from where sexual satisfaction can be elicited; there are so many that we can almost say that there is no part of the female body which does not give sexual response, the partner has only to find the erotogenic zones."

Okay, there are many spots. But is there a special spot? Dr. Grafenberg thought so, "*An erotic zone always could be demonstrated on the anterior wall of the vagina along the course of the urethra* [*the emphasis here is mine*]. Even when there was a good response in the entire vagina, this particular area was more easily stimulated by the finger than the other areas of the vagina. Women tested this way always knew when the finger slipped from the urethra by the impairment of their sexual stimulation. During orgasm this area is pressed downwards against the finger like a small cyst protruding into the vaginal canal. It looked as if the erotogenic part of the anterior vaginal wall tried to bring itself in closest contact with the finger. It could be found in all women, far more frequently than the spastic contractions of the pelvic floor. . . . After the orgasm was achieved a complete relaxation of the anterior vaginal wall sets in."

Similar to the male urethra, the female urethra seems to be surrounded by erectile tissues, called the corpora cavernosa in males. In the course of sexual stimulation, the female urethra begins to enlarge and can be felt easily. It swells out greatly at the end of orgasm. The most stimulating part is located where the urethra arises from the neck of the bladder.

Erotogenic zones in the female urethra are sometimes the cause of urethral masturbation. I have seen several teenage girls with this problem. Girls stimulated themselves with hair pins, pens, or pencils in their urethra. The blunt part of the old-fashioned hair pin, or other small, narrow object, is introduced into the urethra and moved back and forth. During the ecstasy of the orgasm some girls lose control of the object and it ends up in

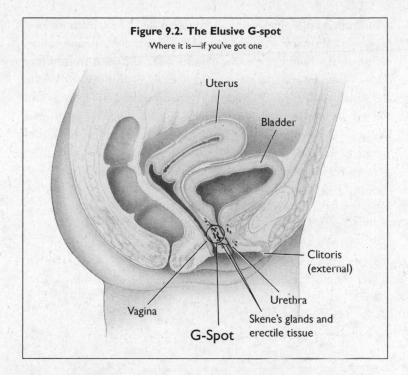

Figure 9.2. The Elusive G-spot

Where it is—if you've got one

Uterus

Bladder

Clitoris
(external)

Urethra

Skene's glands and
erectile tissue

Vagina

G-Spot

the bladder. Most girls who lose these objects feel ashamed and hide the incident from their parents until a bladder problem results. These lost objects have caused bladder stones, infections, and tears. Urethral masturbation also happens in boys, men, and couples. I have seen patients with drink stirrers lodged in their urethras. This happens more when the couple is drunk and they have low prefrontal cortex activity.

Another interesting discovery associated with G-spot research is the incidence of female ejaculation. Milan Zaviacic and his colleagues at Comenius University in Bratislava reported research on this phenomena in an article published in the *Journal of Sex Research*. The study was conducted with twenty-seven women; a G-spot was found in all, and ten of them experienced episodes of feminine ejaculation. The accepted theory is that the G-spot is analogous to the prostate and associated glands, as it seems to produce a fluid chemically similar to male seminal fluid. This phenomenon is one that many women mistake for urinary incontinence, although the fluid released is actually quite different from urine. This is not an uncommon occurrence, as is shown in "Female Ejaculation: Perceived Origins, the Grafenberg Spot/Area, and Sexual Responsiveness," an article in the *Archives of Sexual Behavior*. According to the introduction, 40 percent of the participants in an anonymous mail survey of 2,350 women (with a 55 percent response rate) reported experiences of ejaculation.

Lesson #9: Use brain sex tricks to optimize your love life.

MAKE IT LAST,
MAKE IT UNFORGETTABLE
—
Embedding Yourself in Your Partner's Memory

"It is not the number of breaths you take that matter, but the moments that take your breath away."

—BUDDHIST WISDOM

The brain works through association. It has massive memory banks that tie together every idea, thought, feeling, action, image, smell, touch, taste, and sound. Every moment we are alive we are a combination of our present state and past memories put together. Every moment you are together with your partner, and many moments throughout the day when you are apart, you are embedding yourself in your partner's memory circuits. If you want to be a positive force in your partner's brain, it is important to plant the seeds of excitement, happiness, novelty, and joy, rather than boredom, anger, or insecurity. In this chapter I'll discuss twelve brainy ways to embed yourself into your partner's brain to make love last and to make yourself unforgettable.

A word of caution: Do these things only if you are sincere and want an emotional and neurochemical attachment to the other person. When people get tightly bonded to others, which likely will happen if you follow the suggestions below, it is very hard for them to separate or break up with their loved one. Relationship breakdowns are horrible when bits of the other person are stored in every fun part of your brain. The grieving process can be very hard on your immune system. People are known to become sick and even die when an intense relationship ends. Giving someone hundreds of cards and then abandoning him or her is cruel and likely to cause a person to feel depressed, obsessed, and perhaps even to set up stalking behavior. If you are going to connect yourself to others, be serious. Otherwise, get a dog, get a hooker, or get a vibrator.

Twelve Ways to Make Your Love Unforgettable

1. Take your partner's breath away. Do something amazingly thoughtful and out of the ordinary. These events solidify you in the person's limbic brain. A close friend was dating a new woman. On his birthday, his girlfriend gave him birthday cards signed by all four of his siblings and his ten cousins. He was stunned by the thoughtfulness of the gesture. Not only was it unique, it showed that she thought about him and planned something special weeks ahead of time. Her thoughtfulness was embedded in his memory.

Once I gave a beautiful flower arrangement to my sweetheart. I knew it would be special, because five women who saw the arrangement at the store asked if I would be their boyfriend. Flowers are one of the best brain gifts. The scent helps to soothe and activate the limbic brain. The smells last for days and they have no calories to make someone fat. My sweetheart was grateful for weeks. Along the same lines, once I received a huge flower arrangement from a woman I was dating. It so surprised me that it embedded her kindness in my heart for a long time. Men love getting flowers, too.

The thoughtfulness does not have to be about money. It means the most when it is about the time spent and forethought that goes into making it happen. In a similar vein, writing poetry or making someone a sentimental CD can also be very special.

Taking someone's breath away involves surprise, in a wonderful way, even if it is a small gesture of love. You can also teach your partner how you want to be surprised. If you like chocolate, tell him to hide some around the house in creative places so that you can find them throughout the day and think of him. If you like flowers, tell him or her to send them on occasion.

I have a friend who surprised her husband by showing up at his office in a trench coat, with nothing on underneath except fishnet stockings and sexy shoes. This was so much fun for both of them. Another wonderful little surprise is to leave notes in pockets so that he or she can find them throughout the day. Another idea is to pick up your partner from work for a surprise gourmet lunch in a beautiful park. Find out what makes your partner tick and then find a way to tie it into your little plot to bring more joy and pleasure into his or her world.

2. Do something special on a regular basis. One of the best ways to make yourself unforgettable is to do things for your partner on a regular basis. Make his or her nervous system expect your call, want to hear your voice, miss your touch or the look into your eyes. Giving someone greeting cards, paper cards, or e-cards on a regular basis is a wonderful way to stay connected. Being your partner's first call in the morning and last call at night helps to solidify you in his or her neural networks. Many greeting card companies and florists have programs to remind you on a regular basis to send something special. It is the reinforcement of thoughtfulness that makes a difference.

3. Do something special on an intermittent or unpredictable basis. This may even be more powerful than doing something regularly. Sometimes, when we do something on a regular schedule, people come to expect it and it loses its power, sort of like living at the beach and seeing sunsets every night. They tend, for some (not me), to lose their special nature. There is a concept in "learning theory" called intermittent reinforcement. It is one of the most powerful learning tools known. Here, the reinforcers, such as

flowers, are not sent on a regular schedule, but rather on an irregular one, such as once or twice a month. The trick is to make it unpredictable and unexpected.

4. Frequent, loving eye contact (some cultures call it eye gazing) is an especially powerful connection tool for bonding. Eye contact enhances intimacy. No eye contact decreases bonding and connectedness. New York psychologist Professor Arthur Arun has been studying the dynamics of what happens when people fall in love. He has shown that the simple act of staring into each other's eyes has a powerful impact. He asked two strangers to reveal intimate details about their lives to each other. They did this for an hour and a half. The two strangers were then made to stare into each other's eyes without talking for four minutes. Afterward, many of his couples confessed to feeling deeply attracted to their opposite number, and two of his subjects even married later. When we are aroused and interested in what we are looking at, our pupils dilate. In medieval Italy, women put the chemical belladonna (literally meaning "beautiful lady") into their eyes to make their pupils look bigger. However, this is not recommended, as belladonna can be poisonous and eventually leads to blindness. It is better and safer to stick to eyeliner and mascara.

5. Learn what pleases your partner sexually. Their pleasure should be your pleasure, if you want to make your love life unforgettable. Gain skill in the things that make him or her happy, in what turns him or her on, in what brings joy. Making this a priority will give you many, many dividends. Think of it as a class (yes, there should be tests) on the ins and outs of your partner. This is one of the best ways to keep the relationship young and happy. As part of the class, you need to know all about pleasing your partner. It shouldn't be a mystery or only talked about when you feel frustrated. It is time for a great collaboration. Here is one exercise I use with couples. Do it on each other. I like to have couples show the other person what they like. Make it a kinesthetic exercise. Show your partner how you like to have your ears kissed, by kiss-

ing her ears the way you would like to be kissed. Show your partner how you like your back being rubbed by rubbing his back the way you like.

Many couples are embarrassed to ask for what they want and what they like. Your partner is probably *not* good at reading your mind. Tell him or her what turns you on. Go into detail. Have your partner practice on you. Try this exercise by filling out the following form and working through each question with your partner.

1. I like it when you _____ with my hair.
2. I like it when you _____ with my ears.
3. I like it when you _____ with my eyes.
4. I like it when you _____ with my nose.
5. I like it when you _____ with my neck.
6. I like it when you _____ with my upper back.
7. I like it when you _____ with my lower back.
8. I like it when you _____ with my breast/chest.
9. I like it when you _____ with my belly.
10. I like it when you _____ with my genitals.
11. I like it when you _____ with my butt.
12. I like it when you _____ with my thighs.
13. I like it when you _____ with my lower legs.
14. I like it when you _____ with my feet.

6. Teach your partner what you like. Most people get joy by pleasing others. Be an expert communicator by sharing your wants and desires. The brain loves the sounds of excitement. Make sure when your partner is pleasing you that you let him or her know.

When Dr. Irwin Goldstein, an expert in sexual medicine, presented recent research findings to a scientific meeting, he said: "It is rare for me to stand in front of an audience and say, 'This is a manuscript that has changed my life.' But this one has done that." The study, published in the *Journal of Sexual Medicine,* seemed obvious. The results showed that females in committed relationships with men who were treated with an impotence drug (Levitra) had better sex. But, the women didn't just like sex better, they liked it better because their bodies worked better. Lubrication was better. Orgasms were more intense. They lusted more. The women's bodies reacted as if *they* were receiving the drug. So a drug they didn't even take affected their bodies. "Her physiology is linked to him," Goldstein says. "Men share problems with women, and the solutions. . . . It totally intrigues me. I can change someone's physiology without treating them. It's the wildest thing!" In fact, the better a man's response to the drug, the better her response to him.

"Entanglement" is a physics concept. Subatomic particles have "partners"—other subatomic particles—with which they can be entangled, sometimes over great distances. If you change one particle, the change affects the other one with which it is entangled. Dr. Goldstein's study is a strong indicator that humans can be entangled. We change when we fall in love, we become one unit, at least sexually. The two shall become one, as it says in the Bible. "There are no other physiologic abilities of men and women that are shared, and that is what is so fascinating about these data," Goldstein says. He also says there is some evidence that when he successfully treats women who suffer from dyspareunia, or pain with intercourse, their men get better erections and have more sexual satisfaction. And he suspects that male partners of women with low libido have poorer erections and that if those women could be treated, the men would improve, too. Taking care of yourself is taking care of your relationship.

7. Sexual novelty can boost lasting love. Some people like routine. It makes them feel safe and comfortable. Others need variation and new challenges. According to Emory University psychiatrist and researcher Gregory Berns, novelty is a central factor in achieving and maintaining satisfaction in life and in intimate relationships. He thinks the basal ganglia, which uses the neurotransmitter dopamine, hold the key. He published his findings in a book titled *Satisfaction: The Science of Finding True Fulfillment.* He starts his exploration with a simple question, "What do humans want?" He

challenges the belief that we are driven primarily to pursue pleasure and avoid pain. Rather, Berns finds that "satisfaction comes
less from the attainment of a goal and more in what you must do
to get there." With a series of experiments using cutting-edge functional MRI scanning technology, he sees that the interaction of
dopamine, the neurotransmitter secreted in the brain in anticipation of pleasure, and cortisol, the chemical released when we are
under stress, produces the feelings people associate with satisfaction. Berns ventures into the world to demonstrate his ideas, studying the bruised and reddened S&M players, as well as looking at
the ultramarathoners who collapse after a hundred-mile run. Berns
then brings his journey home, looking at issues in his own marriage
and the sexual dissatisfaction that so often plagues long-term relationships. His conclusion is simple and compelling: People are
wired for novel experience, and when we seek it out, we are satisfied. Look for ways to bring novelty into your intimate relationships. Do things differently, from the way you kiss your partner, to
the way you show love, to the activities you do together.

8. Do something a little edgy. Along the same lines as novelty,
getting your partners heart rate up may make her more interested
in you. She might interpret the rush as a feeling of excitement for
you. One experiment showed that if people experience fear on a
date, they often misinterpret that feeling as love. So dates at a
theme park may be more successful than a science lecture (unless
it is on the neuroscience of sex). A bungee jump might even seal
your relationship for life. If you do this with an anxious partner,
however, he or she may leave you behind forever to avoid the fear.
It is important to individualize these suggestions to your partner.
For example, being a passenger with someone who drives fast may
be exciting for partners who need excitement and speed, but it
may be a disaster for someone who is naturally more cautious.

9. Use every sense. Utilize all of your partner's senses to make
yourself unforgettable. Our five senses are the vehicles that bring
the outside world in. They are what sees, hears, tastes, smells, and

feels so that we can know another person. A large portion of the brain is dedicated to your senses. It has been estimated, for example, that 50 percent of the brain is dedicated to vision. Use these senses to embed yourself deeply into the sensory circuits of your partner's brain. Embedding works best if you are helping to encode good memories or feelings, rather than annoying ones. Begin by taking a look at how you emit each sense and see if there is a way to smooth out any rough edges to make yourself more desirable, more enticing. For instance, it often amazes me when I hear couples talking to each other. So often the words and voice tone are so grating that I am sure they are embedding anger and irritation day in and day out.

One way to assess your own voice is to listen to it on a tape recorder. The sound of your voice can be a powerful elixir of passion or an irritant to those around you. I used to be surprised when I heard my recorded voice. It sounded different being played back to me than it did when I heard it while I was talking. Record your voice and listen to the quality of it. Then, focus on adding a bit of richness or lower tonal quality to your voice and record it again. When you play it back, hear if your voice sounds more pleasing to you. If it does, keep working on it. Initially, it takes practice to be conscious of such aspects of your self, such as the voice, but with time it gets easier. Some people benefit from a voice coach to help smooth it out. When I started to do radio and television interviews, I used a wonderful voice coach who was ever so helpful. When a friend from Buffalo decided to lose some of her accent to do more voiceover work, it took her almost a year of conscious speaking before it became her natural sweet voice. I'm not suggesting that anyone change who he is to be more appealing, rather than to augment the natural beauty that already exists within. Just because we develop a habit or certain way of doing something doesn't mean that it's the best way to do it. So, be open to using the natural talents that exist within you to be more of who you truly are in the world around you.

Once you have found a nice tonal quality and voice level and speed, leave a sultry message for your mate on voice mail. By leaving a recorded message that is fun and appealing, your mate can play it back, over and over, whenever he or she needs a lift or a smile. This is a wonderful way of embedding yourself in your partner's day-to-day life.

Next, look at the other aspects of your senses to see if they can be refined for the purpose of embedding yourself in your partner's consciousness. For instance, it's nice to take some time to get a manicure and pedicure, shave and groom a few extra moments, check your breath, keep your lips soft and supple for more sensual kisses, find a scent that your partner likes and wear it to embed your odor on clothing that he or she will smell throughout the day. If you like to cook or draw, make something special and send it along on his or her journey for the day. Take pictures of yourself, as well as yourself and your partner, in clothing or environments that would inspire your mate while looking at it. Frame that picture and give it as a gift or place it somewhere where it will be seen often. In feng shui, it is believed that a loved one's picture placed in view of the bed is good for that relationship. Flowers or little knickknacks that your mate would like as gifts are nice reminders that you are near, as these items can be placed within immediate sight. Use every sense available, be it smell, sight, sound, taste or touch.

Here are more sensory specific ideas:

Vision. Give pictures of yourself and your loved one embracing. Wear sexy clothes. Go shopping with your partner and let him or her help you pick out part of your wardrobe. Wear clothes that he or she likes on you. Take care of your appearance. Find out if there are any visual cues that upset your partner and take care of them if possible. Add candles or a fire to the environment. The light from gentle flames often enhances sensuality.

Sound. Watch the sound and tone of your voice. Decrease distractions when you are together. Music is a wonderful way to connect with your partner. By introducing your mate to new sounds and choosing "your song" together, a switch will be flipped to remind her of you whenever she hears it. This can be a double-edged sword if you break up. One woman I dated once made me a fabulous mix CD of my favorite songs and a few of hers. I was so touched by this gesture that every time I played the CD I had kind thoughts of her. Be conscious of your partner's taste in music and when a new album comes out or that band comes to town that he really likes, buy tickets and experience his favorite sounds together. Learn to dance together to this music. Perhaps, even create exotic dances that you can do for your partner to melodies that are really impassioning to them. The subconscious power of music is one of the most powerful elixirs of passion that can be utilized to strengthen a long-term partnership.

Smell. Put sweet fragrances in the environment, such as flowers or potpourri. Cook his or her favorite foods. Use some of the aphrodisiac foods and scents from Lesson Nine. Take a shower, or not, before you make love. Check with your partner to see what he or she likes best. Napoleon specifically asked Josephine not to take baths for two weeks before he came home from battle; he loved her natural scent.

Touch. Touch is very important in the embedding process. Learn to touch your partner in subtle, pleasing ways, be it massaging while she is driving or working at the computer, placing your hand on her while you are talking, brushing your head against his body as you walk by, kissing her in places that are pleasing to her throughout the day, learning to massage him in ways that create ecstasy, and learning to sleep so that some part of you and your partner is in contact. Every sense is an entry point to embed yourself in your partner's mind. If you take the time to do this practice, it will make all the difference in the world in your intimate encounters and longevity of your relationship.

Kisses belong under touch and taste. Remember the very

strange-looking little man, the homunculus, that represented the sensory strip in the brain that processes touch? He has very large lips, because there are millions of nerve endings in the brain dedicated to lips. Kisses are critical to bonding. Do you have lips that fit your partner's? Learn to use your lips in ways that please him or her. Pay attention (use your prefrontal cortex) when kissing. What stimulates him or her? What causes her to moan and come back for more? Linger when you kiss. I once read a book by Ellen Kreidman called *The 10 Second Kiss*. I thought it was a brilliant embedding technique. Her premise was that rather than giving a quick peck to your partner to say hello or good-bye, spend ten seconds and make the moment last. She says that a "10-second kiss" can transform a simple act into a clear message of "I love you" each and every day. The technique takes communication to a level so deep, it's magical. Eliminate pecks; we are not chickens.

Taste. Avoid bad tastes and enhance great ones. Kissing someone with the taste of candy or cinnamon on their lips can be very sexy. Take care of bad breath, but see what tastes your partner loves. Someone may really think Listerine mouthwash is the sweet elixir of passion, while others might feel it reminds them of going into surgery. Check out what your partner likes. Also, keep the genital area as tasty as possible. Some people love performing oral sex on their partner after they have bathed, others prefer a more natural state. Ask. Feeding each other during meals can also be very exciting. When you taste something you love, feeding it to your partner with words like, "This is great, try it," shows her that her pleasure is on your mind.

10. Do something great for someone your partner loves. When we take care of important people in our partner's life, we take care of him or her. This is one of the most powerful bonding techniques I know. When you care for your partner's children, parents, friends, employees, or even pets, the partner's limbic brain is grateful and you become more deeply embedded in his or her consciousness.

I am often amazed at how many people do not understand this technique. Given that there are many second and third marriages in our society, dealing with a partner's children is often an important issue. Setting up competition or conflict between a new love and his or her children almost always leads to disaster. In bonding to someone, spend time with his or her children, or help her parents and you will be much more likely to be solidified in her head.

11. Summarize and immortalize loving moments. When you have great moments with your lover, write them down and send your missive to him or her. When you experience a great moment, that by itself helps to embed it into memory; when you then take the time to write it down, it helps to further engrain it into the memory tracks of the emotional brain.

Writing loving thoughts have occurred from the beginning of recorded history.

> Bridegroom, dear to my heart,
> Goodly is your beauty, honeysweet. . . .
> Bridegroom, I would be taken by you to the bedchamber.
> You have captivated me,
> Let me stand tremblingly before you.

That's the enticing start to the oldest love known poem in the world. Scholars discovered the poem over a hundred years ago buried in the ancient sands of Iraq. The poem was written around 2030 BC by a Sumerian scribe from the city of Ur using a reed stylus on wet clay, which was then baked, preserving the tablet of passion for forty centuries. The passion, scholars say, was part of a Mesopotamian festival ritual of fertility and power called Sacred Marriage.

12. Learn from parrots. Neurologist Barbara Wilson has trained and kept parrots for years. She says that they have taught her a lot about relationships (no kidding): Share your food with the one that you love, groom each other, sing constantly, build

nests together, *and repeat* each other's words and actions! If people would think like parrots, all babies would be planned. Parrots don't just randomly mate. First, they have to genuinely like the other bird. Then there has to be a constant reliable source of food, light, and stability. Then there has to be a nest box. The male inspects it first, then the female. The female waits until the male feeds her. Only then do they get frisky. If one dies, the other one in the mated pair grieves and mourns and has been known to just up and die, too.

Barbara once had a mealy Amazon parrot that she rescued from a too small cage at a breeder's farm. Having been stolen from the wild (you can trace their import origin from the type and numbering on the band on their leg), the female bird had not seen another mealy (specific subspecies of parrot with its own squawking language) since its infancy. She was depressed and overweight from being in the confining cage. Barbara put the bird in a large cage, where the parrot got significantly better, but still she was lonely. Barbara never heard a peep out of the bird, even when she once cut her toenail too short. Barbara did some Internet parrot dating for the bird and found a single male of her subspecies in south Texas. After several e-mails, phone calls, and high-level bird trading, Barbara drove sixteen hours round-trip to get her parrot a "man." When she brought the male back, both birds were ecstatic. They squawked for hours at loud volume. The two birds *dated* for three weeks. They would not share a cage. Barbara opened the two cages, they would meet on top, and squawk for hours. Eventually they groomed each other, then each night they would go back to their own cage. Finally, Barbara says she had a talk with them on the human facts of life: She only had room for one big cage and was tired of cleaning two cages, so like it or not, they were going to have to live together. They slept on separate perches for three more months despite genuinely loving each other. The female lost weight and looked years younger due to the affection. Then they slept together, ate together, and made a life that worked.

When you ask her about parrot behavior, she says, "It's just your own behavior and words repeated back to you. And if you ignore them they start to bite and pick their feathers, and you wonder why you have a mean and ugly bird." After she achieved success with parrots (she had a flock of twelve at one time), she decided it was time to find a husband. Really cool fact, she tells me, men catch on quicker than birds.

Lesson #10: Make it last by leaving enduring impressions on your partner's brain.

FIX THE BRAIN ISSUES
THAT GET IN THE WAY OF SEX
——
PMS, Depression, ADD, Substance Abuse,
Denial, and Being a Jerk

Get away, get away, get away, get away
Get away cause I'm pms-ing

—"PMS," Mary J. Blige

Celia and Greg fell madly in love for five weeks. They met on Match.com shortly after Labor Day. On the surface they seemed like a perfect couple. They were both well educated, caring, hardworking, and had similar lifestyle habits. The attraction was amazing, even between their families. Greg loved Celia's little girl, and Greg's teenage girls got along very well with Celia. With new love comes hope. They were together most days and on the phone for several hours a day when they were apart. Five weeks into the passionate relationship things started to abruptly change. Celia started to back up. She became distant, irritable, and short tempered. Nothing Greg did seemed right. Even though she was fully involved with moving the relationship forward, she felt the need to backtrack. Greg felt disoriented. What happened?, he wondered. Initially, he felt anxious at the change. He had met few women as wonderful as Celia, few women that took his breath away. But he did as Celia wanted. That still didn't seem good enough and Celia broke off the relationship. Greg felt very sad. Then Celia started her menstrual period. She was horrified by her behavior and for losing Greg. When she called him, he was happy to hear from her, but hurting from what happened. He was gun shy. Premenstrual tension syndrome (PMS) is real and causes real problems in the brain and in relationships.

It is not just PMS that can ruin relationships. Other brain problems, such as depression, ADD, substance abuse, anxiety disorders, obsessive-compulsive disorder, and personality disorders also interfere with love. Understanding and treating these problems is critical to healthy relationships and healthy sex. In this chapter, I will explore the most common brain ailments interfering with love and sex that we see in our clinics and give you a way to think about how to get the best help for them. Some people will need psychotherapy; some will need medication; others will need more directed guidance with supplements or other alternative treatments. I will also help you decide if and when you need to seek professional help. In lecturing around the world, I am frequently asked the following questions: When is it time to see a professional about my brain? What should I do when a loved one is in denial about needing help? How do I go about finding a competent professional?

A Quick View of Common Brain Problems Affecting Love and Sex

PMS

When I saw patients with PMS after I started my brain-imaging work in 1991, I just had to look. Now I know more about PMS than I want to. I have five sisters and two daughters. Plus, I have an ex-lover (I'll call her Laura) who suffered from severe PMS. She loved me passionately for the first seven days of her menstrual cycle, was very neutral on me for the next fourteen days, and just seemed to hate me for the last seven days or so of her cycle.

Laura's behavior the first seven days of her cycle kept me hooked into the relationship. Our relationship was being *intermittently reinforced,* a psychological term about learning behavior; when someone is reinforced occasionally or intermittently, it causes them to want to stay in a relationship, hoping for more.

Over the past years we have scanned many women with PMS just before the onset of their period, during the worst time of their cycle, and then again a week after the onset of their period, during the best time. Even though brain-SPECT scans are very consistent from day to day in most people, they can radically change in women with PMS. I knew from my own experience with Laura that likely the PMS brain changed over the month. When PMS is present, we see dramatic differences between the scans. When a woman feels good, her deep limbic system (emotional brain) is calm and cool and she has good activity in her temporal lobes (mood stability and memory) and prefrontal cortex (judgment). Right before her period, when she feels the worst, her deep limbic system and anterior cingulate gyrus (worry center) is often overactive and she has poor activity in her temporal lobes and prefrontal cortex!

I have seen two PMS patterns, clinically, and on SPECT, that respond to different treatments. One pattern is increased deep limbic activity often accompanied by poor activity in the temporal lobes, which correlates with cyclic mood changes and anger. This finding often responds best to anticonvulsant medications, such as Depakote, Neurontin, Lamictal, or Tegretol. These medications tend to even out moods, calm inner tension, decrease irritability, and help people feel more comfortable in their own skin.

The second PMS pattern that I have noted is increased deep limbic activity in conjunction with increased anterior cingulate gyrus activity. The anterior cingulate, as we have seen, is the part of the brain associated with shifting attention. Women with this pattern often complain of increased sadness, worrying, repetitive negative thoughts and verbalizations (nagging), and cognitive inflexibility. This pattern usually responds much better to medications that enhance serotonin availability in the brain, such as Lexapro, Zoloft, and Prozac. Here are two examples.

Brittany. Brittany was a thirty-eight-year-old married female referred for evaluation of suicidal thoughts, depression, and temper flares. She also experienced problems with anxiety, excessive tension, and overeating. These problems occurred primarily during the last ten days of her menstrual cycle and abated two to three days after the onset of menses. On several occasions she separated from her husband within the seven days prior to the onset of her period; on one occasion, she lashed out at him physically. The patient and her husband confirmed the cyclic changes to her symptoms. Both Brittany and her husband kept a symptom log over the next month. On Day 27 (of a twenty-nine-day cycle) Brittany called the clinic saying that she was having problems with suicidal thoughts and depression. She was scanned the same day. Her SPECT study revealed significant increased activity in the anterior cingulate gyrus and marked decreased activity in the left temporal lobe and prefrontal cortex bilaterally. She was then scanned on Day 8 of the next menstrual cycle when she was symptom free. Her follow-up scan revealed improved temporal lobe and prefrontal cortex function but persistent cingulate hyperactivity. Due to the clear temporal lobe problems, Brittany was placed on the anticonvulsant Depakote, which stabilized her temper outbursts and suicidal thoughts. The serotonergic antidepressant Zoloft was then added a month later due to persistent premenstrual sadness. Three years later she remains symptom free.

Anne. Anne was a thirty-three-year-old married female referred for evaluation of suicidal thoughts, depression, anxiety, and irritability. These problems occurred predominantly during the last week of her menstrual cycle and significantly let up several days after the onset of menses. She had experienced a postpartum depression after the birth of one child but not after the birth of her other two children. Anne and her husband confirmed the cyclic changes to her symptoms. Both she and her husband kept a symptom log over the next month. On Day 25 (of a twenty-eight-day cycle) Anne called the clinic complaining of severe agitation and moodiness. She was scanned the same day. Her SPECT study revealed significant increased activity in the anterior cingulate gyrus and deep limbic regions. She was then scanned on Day 10 of the next menstrual cycle when she was symptom free. Her follow-up scan revealed excess activity in the anterior cingulate gyrus and deep limbic system. Lexapro was very effective in calming her symptoms. Two years later she remains symptom free during the premenstrual period.

Mood Disorders

Mood disorders severely affect libido and relationships. Depression is often associated with low libido, negativity, and a higher divorce rate. The "up" or manic phase of bipolar disorder can be associated with impulsivity, hypersexuality, and hyperreligiosity.

Depression

Burl, a fifty-two-year-old contractor, husband and father of two boys, was referred to me because he was tired all the time. His family physician ruled out the physical causes of fatigue and thought he was stressed. Additionally, he had trouble focusing at work and had trouble sleeping. His caffeine use went way up, but it didn't help his energy, just made him edgy. His sex drive was gone, his appetite was poor, and he had no interest in doing things with his family. Burl would cry for no apparent reason and he even began to entertain suicidal thoughts. Burl had a serious depressive illness.

Depression is a very common brain illness. Studies reveal that at any point in time, 3 to 6 percent of the population have a significant depression. Only 20 to 25 percent of these people ever seek help. This is unfortunate because depression is a very treatable problem.

The following is a list of symptoms commonly associated with depression:

- sad, blue, or gloomy mood
- low energy, frequent fatigue
- lack of ability to feel pleasure in usually pleasurable activities
- irritability
- poor concentration, distractibility, poor memory
- suicidal thoughts, feelings of meaninglessness
- feelings of hopelessness, helplessness, guilt, and worthlessness
- changes in sleep, either poor sleep with frequent awakenings or increased sleep
- changes in appetite, either markedly decreased or increased
- social withdrawal
- low self-esteem.

Early detection and treatment is important to a full and complete recovery. My imaging work has revealed that there are multiple types of depression and treatment needs to be specifically tailored to the type. See my book *Healing Anxiety and Depression* (written with Lisa Routh).

Bipolar Disorder

Patricia is a twenty-eight-year-old married mother of two children. She had a period of depression six months earlier and had been prescribed an antidepressant by her OB/GYN. Initially she felt much better. Then she started slowly having trouble sleeping. Her thoughts raced, she became more irritable and much more sexual. She was used to having sex several times a week with her husband, but now wanted it every day. She propositioned three of her male coworkers, which was out of character. Two of her coworkers took her up on her offer and she ended up contracting herpes, which she gave to her husband. On the verge of divorce, they came to see me. Patricia had bipolar disorder triggered by the antidepressant, which is not an uncommon scenario. It is sad to think that an improperly treated psychiatric illness can tear apart families. With the right treatment, which included a mood stabilizer and fish oil, Patricia and her husband did much better.

Bipolar disorder is a mood illness where people cycle between two poles of emotion. There may be periods of depression that alternate with periods of high, manic, irritable, or elated moods. Mania is categorized as a state distinct from one's normal self, where there is greater energy, racing thoughts, impulsivity, a decreased need for sleep, and a sense of grandiosity. It is often associated with periods of hypersexuality, hyperreligiosity, or spending sprees. Sometimes it is also associated with hallucinations or delusions. In treating the depressive part of the cycle, both pharmaceutical and supplemental antidepressants have been known to stimulate manic episodes. It is important to vigorously treat this disorder, as it has been associated with marital problems, substance abuse, and suicide.

Here is a list of symptoms often associated with bipolar disorder:

1. Periods of abnormally elevated, depressed, or anxious mood
2. Periods of decreased need for sleep, feeling energetic on dramatically less sleep than usual
3. Periods of grandiose notions, ideas, or plans
4. Periods of increased talking or pressured speech
5. Periods of too many thoughts racing through the mind
6. Periods of markedly increased energy
7. Periods of poor judgment leading to risk-taking behavior (separate from usual behavior)
8. Periods of inappropriate social behavior
9. Periods of irritability or aggression
10. Periods of delusional or psychotic thinking.

Bipolar I, which used to be called manic depressive illness, is thought to be the more classic form of this disorder. In recent years, a milder form of the disorder, called Bipolar II, has been described; it is associated with depressive episodes and milder "hypomanic" issues.

The treatment for bipolar disorder, both I and II, is usually medication, such as lithium or anticonvulsants such as Depakote. Recent literature suggests that high doses of omega-3 fatty acids, found in fish or flaxseed oil, can also be helpful.

Antidepressants and Romantic Love

Depression inhibits romantic love. People who are depressed tend to be negative, socially isolated, and often have little interest in sex. They might also have suicidal feelings, which are usually a turnoff to potential partners. Treating depression is essential for people to have healthy relationships. Yet, the specific type of treatment can either enhance or hurt attraction in romantic love. In my imaging work I have discovered that depression is not one illness and that the treatment needs to be tailored individually. In general, however, bupropion (Wellbutrin) is prosexual and enhances sexual feelings and function. It enhances dopamine availability in the brain and increases attention and focus. It is useful to treat depressions associated with low energy. SSRIs, useful in treating depression associated with obsessive thinking, can jeopardize romantic love. Low serotonin levels help explain the obsessive thinking common in early romantic love. In a study by Helen Fisher, subjects reported that they thought about their loved one 95 percent of the day and couldn't stop thinking about them. This kind of obsessive thinking is comparable to obsessive-compulsive disorder, also characterized by low serotonin levels. Serotonin-enhancing antidepressants blunt the emotions, including the elation of romance, and suppress obsessive thinking, a critical component of romance. "When you inhibit this brain system," Dr. Fisher warns, "you can inhibit your patient's well-being and possibly their genetic future." These antidepressants also inhibit orgasm, clitoral stimulation, penile erection, and seminal fluid. From an anthropological perspective, Dr. Fisher concludes, "a woman who can't get an orgasm may fail to distinguish Mr. Right from Mr. Wrong." As one woman on an SSRI told me, "I thought I no longer was attracted to my husband." In a study, women on SSRIs rated male faces as more unattractive, a process called courtship blunting. Seminal fluid contains dopamine and norepinephrine, oxytocin and vasopressin, testosterone and estrogen. Without an orgasm, men may lose the ability to send courtship signals. These warnings should encourage us to look for alternative treatments in depression.

Anxiety Disorders

There are five common types of anxiety disorders that can affect people's relationships, moods, and sexuality in a negative way: panic disorders, agoraphobia, obsessive-compulsive disorder, posttraumatic stress disorder, and performance anxiety. I'll briefly discuss each of these and their treatments.

Panic Disorder

Healthy sexuality is usually enhanced by a sense of safety and peacefulness. But what if all of a sudden your heart starts to pound. You get this feeling of incredible dread. Your breathing rate goes faster. You start to sweat. Your muscles get tight, and your hands feel like ice. Your mind starts to race about every terrible thing that could possibly happen and you feel as though you're going to lose your mind if you don't get out of the current situation. You've just had a panic attack. Panic attacks are one of the most common brain disorders. It is estimated that 6 to 7 percent of adults will at some point in their lives suffer from recurrent panic attacks. They often begin in late adolescence or early adulthood but may spontaneously occur later in life. If a person has three attacks within a three-week period, doctors make a diagnosis of a panic disorder.

In a typical panic attack, a person has at least four of the following twelve symptoms: shortness of breath, heart pounding, chest pain, choking or smothering feelings, dizziness, tingling of hands or feet, feeling unreal, hot or cold flashes, sweating, faintness, trembling or shaking, and a fear of dying or going crazy. When the panic attacks first start, many people end up in the emergency room because they think they're having a heart attack. Some people even end up being admitted to the hospital.

Anticipation anxiety is one of the most difficult symptoms for a person who has a panic disorder. These people are often extremely skilled at predicting the worst in situations. In fact, it is often the anticipation of a bad event that brings on a panic attack. For example, you are in the grocery store and worry that you're going to have an anxiety attack and pass out on the floor. Then, you predict, everyone in the store will look at you and laugh. Pretty quickly the symptoms begin. Sometimes a panic disorder can become so severe that a person begins to avoid almost any situation outside of home—a condition called agoraphobia.

Panic attacks can occur for a variety of different reasons. Sometimes they are caused by medical illnesses, such as hyperthyroidism, which is why it's always important to have a physical examination and screening blood work. Sometimes panic attacks can be brought on by excessive caffeine intake or alcohol withdrawal. Hormonal changes also seem to play a role. Panic attacks in women are seen more frequently at the end of their menstrual cycle, after having a baby, or during menopause. Traumatic events from the past that somehow get unconsciously triggered can also precipitate a series of attacks. Commonly, there is a family history of panic attacks, alcohol abuse, or other mental illnesses.

On SPECT scans we often see hyperactivity in the basal ganglia, or sometimes temporal lobe problems. Psychotherapy is my preferred treatment for this disorder and in some studies has been shown to calm basal ganglia activity. Sometimes supplements or medications can also be helpful. Unfortunately the most helpful medications are also addictive, so care is needed.

Agoraphobia

The name *agoraphobia* comes from a Greek word that means "fear of the marketplace." In behavioral terms it means the fear of being alone in public places. The underlying worry is that the person will lose control or become incapacitated and no one will be there to help. People afflicted with this phobia begin to avoid being in crowds, in stores, or on busy streets. They're often afraid of being in tunnels, on bridges, in elevators, or on public transportation. They usually insist that a family member or a friend accompany them when they leave home. If the fear establishes a foothold in the person, it may affect his or her whole life. Normal activities become increasingly restricted as the fears or avoidance behaviors dominate their life.

Agoraphobic symptoms often begin in the late teen years or early twenties, but I've seen them start when a person is in their fifties or sixties. Often, without knowing what is wrong, people will try to medicate themselves with excessive amounts of alcohol or drugs. This illness occurs more frequently in women and many who have it experienced significant separation anxiety as children. Additionally, there may be a history of excessive anxiety, panic attacks, depression, or alcohol abuse in relatives.

Agoraphobia often evolves out of panic attacks that seem to occur "out of the blue," for no apparent reason. These attacks are so frightening that the person begins to avoid any situation that may be in any way associated with the fear. I think these initial panic attacks are often triggered by unconscious events or anxieties from the past. For example, I once treated a patient who had been raped as a teenager in a park late at night. When she was twenty-eight, she had her first panic attack while walking late at night in a park with her husband. It was the park setting late at night that she associated with the fear of being raped and which triggered the panic attack. Agoraphobia is a very frightening illness to the patient and his or her family. With effective, early intervention, however, there is significant hope for recovery. The scan findings and treatment are similar to those for people with panic disorder. The one difference is that people with agoraphobia often have increased anterior cingulate gyrus activity and get stuck in their fear of having more panic attacks. Getting stuck in the fear often prevents them from leaving home. Using medications, such as Prozac and Lexapro, or supplements, such as 5-HTP and St. John's wort, to increase serotonin and calm this part of the brain is often helpful.

Obsessive-Compulsive Disorder (see Lesson Eight)

Posttraumatic Stress Disorder

Joanne, a thirty-four-year-old travel agent, was held up in her office at gunpoint by two men. Four or five times during the robbery, one of the men held a gun to her head and said he was going to kill her. She graphically imagined her brain being splattered with blood against the wall. Near the end of this fifteen-minute ordeal, they made her take off all her clothes. She pictured herself being brutally raped by them. They left without touching her, but locked her in a closet.

Since that time her life had been thrown into turmoil. She felt tense, and was plagued with flashbacks and nightmares of the robbery. Her stomach was in knots and she had a constant headache. Whenever she went out, she felt panicky. She was frustrated that she could not calm her body: her heart raced, she was short of breath, and her hands were constantly cold and sweaty. She hated how she felt and she was angry about how her nice life had turned into a nightmare. What was most upsetting to her were the ways that the robbery affected her marriage and her child. Her baby picked up the tension and was very fussy. Every time she tried to make love with her husband, she began to cry and get images of the men raping her. Joanne had posttraumatic stress disorder (PTSD), a brain reaction to severe traumatic events such as a robbery, rape, car accident, earthquake, tornado, or even a volcanic eruption. Her symptoms are classic for PTSD, especially the flash-backs and nightmares of the event.

The worst symptoms came from the horrible thoughts about what never happened, such as seeing her brain splattered against the wall and being raped. These thoughts were registered in her subconscious as fact, and until she entered treatment she was not able to recognize how much damage they had been doing. For example, when she imagined that she was being raped, a part of her began to believe that she actually was raped. The first time she had her period after the robbery, she began to cry with relief that she was not impregnated by the robbers, even though they never touched her. A part of her even believed she was dead because she had so vividly pictured her own death. A significant portion of her treatment was geared to counteract these erroneous subconscious conclusions.

Without treatment, PTSD can literally ruin a person's life. The most effective treatment is usually psychotherapy. One type of psychotherapy that I think works especially well for PTSD is called eye movement desensitization and reprocessing (EMDR). You can learn more about this technique in my book *Healing Anxiety and Depression* or visiting www.emdria.org. Depending on the severity of PTSD, certain types of medications and supplements can also be helpful.

Performance Anxiety

Feeling anxious or nervous before speaking or performing in front of a group is one of the most common fears of human beings. Many people hate feeling judged, scrutinized, or "on the spot." As anxiety levels go up in the brain, thoughtfulness usually goes down. This is particularly true with sexual performance. It is very common for lovers, especially new lovers, to want to please their partners. They feel anxious and their nerves tend to get in the way of sexual play and enjoyment. Often this type of anxiety is associated with what I call Fortune Telling ANTs. ANT stands for automatic negative thoughts. These are the thoughts that go through your head automatically and ruin your day. Fortune-telling ANTs are the thoughts that predict a bad turnout, even though there is no evidence for the idea. Examples include, "He will not like my body." "She will think I have a small penis." "I will come too quickly," "I have to fake an orgasm, or he will not be happy with me." The problem with fortune-telling ANTs is that your brain makes happen what it sees. If you predict failure, you are more likely to fail. For example, when you see yourself not pleasing your partner, the subsequent anxiety will interfere with your feeling relaxed and present, in the moment; then you will not read his or her body language and end up missing important clues to making it a special sexual time. Learning how to calm performance anxiety, through correcting negative thoughts, deep breathing, and meditation, is essential for great sex.

Attention Deficit Disorder

In my lectures I often ask, "How many people are married? Raise your hands." Usually, a good portion of the audience raises their hands. "Is it helpful," I ask next, "to say everything you think in your marriage?" Everyone laughs. "Of course not," I continue, "relationships require tact, they require thoughtfulness. Saying everything you think is usually a disaster in relationships." Unfortunately, you need healthy PFC activity to suppress the sneaky thoughts that just creep through your brain. Attention-deficit/hyperactivity disorder (ADHD) is usually associated with low PFC activity and people are more likely to blurt out, without forethought.

Do you often feel restless? Have trouble concentrating? Have trouble with impulsiveness, either doing or saying things you wish you hadn't? Do you fail to finish many projects you start? Are you easily bored or quick to anger? If the answer to most of these questions is yes, you might have attention deficit disorder (ADD).

ADD is the most common brain problem in children and

adults, affecting 8 to 10 percent of the United States population. The main symptoms of ADD are a short attention span, distractibility, disorganization, procrastination, and poor internal supervision. It is often, but not always, associated with impulsive behavior and hyperactivity or restlessness. Until recently, most people thought that children outgrew this disorder during their teenage years. For many, this is false. While it is true that the hyperactivity lessens over time, the other symptoms of impulsivity, distractibility, and a short attention span remain for most sufferers into adulthood. Current research shows that 60 to 80 percent of ADD children never fully outgrow this disorder.

Common symptoms of the adult form of ADD include poor organization and planning, procrastination, trouble listening carefully to directions, distractibility, short attention span, relationship problems, and excessive traffic violations. Additionally, people with adult ADD are often late for appointments, frequently misplace things, may be quick to anger, and have poor follow-through. There may also be frequent impulsive job or relationship changes and poor financial management. Substance abuse, especially alcohol or amphetamines and cocaine, and low self-esteem are also common.

Many people do not recognize the seriousness of this disorder and just pass off these kids and adults as lazy, defiant, or willful. Yet, ADD is a serious disorder. Left untreated, it affects a person's self-esteem, social relationships, and ability to learn and work. Several studies have shown that ADD children use twice as many medical services as non-ADD kids, up to 35 percent of untreated ADD teens never finish high school, 52 percent of untreated adults abuse substances, teens and adults with ADD have more traffic accidents, and adults with ADD move four times more than others.

Many adults tell me that when they were children, they were in trouble all the time and had a real sense that there was something very different about them. Even though many of the adults I treat with ADD are very bright, they are frequently frustrated by not living up to their potential.

From our research with SPECT scans, it is clear that ADD is a brain disorder, but not one simple disorder. I have described six different types of ADD. The most common feature of ADD is decreased activity in the prefrontal cortex with a concentration task. This means that the harder a person tries, the less brain activity they have to work with. Many people with ADD self-medicate with stimulants, such as caffeine, nicotine, cocaine, or methamphetamine, to increase activity in the PFC. They also tend to self-medicate with conflict-seeking behavior. If they can get someone upset, it helps to stimulate their brain. Of course, they have no idea they do this behavior. I call it unconscious, brain-driven behavior. But, if you are around ADD people long enough, you will see and feel the conflict-seeking behavior.

The best treatment for ADD depends on the type of ADD a person has. See my book *Healing ADD* for a complete description of types and treatments. In general, intense exercise helps, as does a higher protein, lower carbohydrate diet. Sometimes medications or supplements are helpful, but sometimes they can make things worse if they are not right. When correctly targeted, ADD is a highly treatable disorder.

Being a Jerk or a Bitch

I know "being a jerk" or "being a bitch" are not medical diagnostic terms. They are negative, name-calling phrases about someone's behavior. Yet, in my experience, it is possible that these behaviors are the result of poor brain function and not completely under conscious will. Sometimes, a head injury, toxic exposure, sleep deprivation, and personality disorders (see below) can interfere with someone's effectiveness in social and sexual situations.

Our character can be defined in part by the way we interact with others. When the way in which we interact with others doesn't work, when we notice a pattern of multiple relationships and multiple disconnections, it may be that a personality disorder is at the root of the problem. The term *personality disorder* implies inflexible and long-standing patterns of experience and behavior (Diagnostic and Statistical Manual of Mental Disorders IV, or DSM-IV) that impair healthy functioning. They can be the source of great personal pain for the person suffering and those he or she loves. A personality disorder can sabotage relationships, prevent the realization of desired goals, and impede our moral development and our spiritual and sexual health. When we are preoccupied, for example, with intrusive thoughts, deep fears of abandonment, or feelings of paranoia or superiority, it's hard to reach beyond the self to the expansive concerns of spirit and morality. It's hard to be our best selves. A person with a personality disorder may feel inexplicably apart from a sense of well-being, of closeness to others, and to God. It may be hard for someone with a personality disorder to feel empathy and thus to feel part of a reciprocally loving community. Feelings of isolation and disconnectedness can lead to the sense of life being meaningless and to devaluing one's own individual contributions. This sense of aloneness and lack of purpose can place people with personality disorders at higher risk for suicide. Personality is what we present to the outside world. It is not the true self, which is broader and deeper than the outward-appearance self. When we think of working on aspects of the personality, we think not of correcting flaws but of opening doors to greater joy and connectedness.

Personality disorders have been traditionally resistant to psychotherapy. Traditional psychiatric thought has focused on developmental causes of these disorders rather than brain abnormalities. It has been my experience that many people labeled as personality disordered are really brain disordered. The implications for treatment are immense—do we talk someone through their difficult behaviors or try to change their brain? Probably we need both.

Antisocial Personality Disorder

Antisocial personality disorder is characterized by a long-standing pattern of disregard for the rights of others and may be an extension of conduct disorder seen in adolescence. The likelihood of developing antisocial personality disorder seems increased in young children with conduct disorder and ADD. People with antisocial personality disorder frequently break rules, inhabit prisons, and have constant relationship and work problems. They often get into fights. With little or no empathy, they may steal, destroy property, manipulate or deceive others for their ends. They tend to be impulsive and lacking in forethought. Traditionally, these people are thought of as evil, bad, and sinful. The work of psychologist Adrienne Raine of the University of Southern California has seriously challenged this notion. Dr. Raine found that compared to a group of healthy men, the MRI scans of the men with antisocial personality disorder showed decreased PFC volume. They are likely dealing with less access to the brain part that controls conscience, free will, right and wrong, and good and evil. A fascinating additional finding of Dr. Raine's work was that people with antisocial personality disorder also had slower heart rates than the control group and decreased sweat gland activity. Lower heart rates and sweat gland activity are often associated with low anxiety states (your hands sweat and your heart races when you are anxious). Could this mean that people with this type of difficult temperament do not have enough internal anxiety? Could the PFC be involved with appropriate anxiety? Intriguing questions. For example, most people feel anxious before they do something bad or risky. If I needed money, and got the thought in my head to rob the local grocery store, my next thoughts would be filled with anxiety:

"I don't want to get caught."
"I don't like institutional food."
"I don't want to be thought of as a criminal."
"I could lose my medical license."

The anxiety would prevent me from acting out on the bad thoughts. But, what if, as Dr. Raine's study suggests, I do not have enough anxiety and I get an evil thought in my head like, "Go rob the store"? With poor PFC activity (a lousy internal supervisor with little to no anxiety), I am likely to rob the store without considering all of the consequences to my behavior. There is an interesting treatment implication from this work. Typically, psychiatrists try to help lessen a person's anxiety. Maybe we have it backward for people with antisocial personality disorder; for them we should try to increase their anxiety.

Narcissistic Personality Disorder

People with narcissistic personality disorder believe that they are special and more unique or gifted than other people. They require constant admiration and recognition for their achievements. A sense of entitlement derived from a bolstered sense of superiority may lead people with narcissistic personality disorder to place great demands on others, expecting their needs to be met immediately, regardless of the inconvenience. Although they may appear confident, they may in fact have very low self-esteem, which they attempt to boost by association with others they imagine being as gifted as they. They may seek connections exclusively with those whom they perceive to be as special and form alliances solely to advance their careers or other endeavors. While lacking empathy and ability to listen patiently to others' concerns, a person with narcissistic personality disorder may spend an inordinate amount of time thinking about what others think of her or him. People with narcissistic personality disorder may belittle or be envious of others' achievements and be unwilling to acknowledge contributions others make to their own successes. A person with narcissistic personality disorder may appear to be rude, condescending, and arrogant, criticizing others while being unable to tolerate criticism him- or herself. A person with NPD often seesaws between a depressed mood because of feelings of shame or humiliation and grandiosity. As with other personality disorders, a person with narcissistic personality disorder may suffer from additional problems such as anorexia, substance abuse, anxiety, and depression. People with narcissistic personality disorders may have overactive cingulate systems, disallowing them to see outside themselves and to take a broader perspective. Poor prefrontal lobe activity may cause the lack of empathy so pronounced in this disorder.

A feeling of social connectedness is the basis of a healthy soul and character. Clinging to the notion that you might be better than others, somehow more privileged or entitled, erects barriers between you and the people to whom you want to get close and makes it impossible to empathize with others' needs. Protecting yourself with distancing tactics such as criticism, disinterest in others' problems, belittling others, or refusing to acknowledge their accomplishments makes it tough to develop a sense of security and companionship, of being loved. It's very hard to make moral decisions from this place, from the position of "What I need is most important." Persistent focus on yourself, your appearance, how others see you, and the neediness that accompanies these anxiety-provoking concerns takes you away from your true self, the self that can focus on what you really care about and what you want your life to be about. Because people with narcissistic personalities disorder may sometimes accomplish external goals, it can be hard to discern the reasons for a lack of connectedness and a hollow spiritual life. Identifying with others; being able to be humble, grateful, and kind; to listen; and to truly appreciate others' caring and contributions to your life readies you to receive spiritual learning.

Borderline Personality Disorder

Instability in relationships, impulsivity, and low self-esteem characterize borderline personality disorder. People with borderline personality disorder may quickly switch attitudes toward others, identifications, values, and goals. For example, someone with borderline personality disorder may worship a new friend or lover and then drop him or her quickly, complaining that their new friend wasn't caring enough. Professional goals and interests may change suddenly, as may moods. Highly reactive and impulsive, they may experience periods of extreme irritability, anger, or anxiety. They may engage in self-destructive behaviors such as drinking heavily, driving fast, overspending, bingeing on food, or having unsafe sex. People with borderline personality disorder may feel periods of great emptiness and engage in suicidal or self-mutilating behaviors. Boredom may be intolerable to someone with this disorder, and consequently, he or she may perpetually seek stimulation. Childhood abuse or neglect or the early loss of a parent may be found in family histories of people with this disorder.

The biological underpinnings of borderline personality disorder are complex. People with borderline personalities may have a combination of prefrontal lobe problems, which accounts for impulsivity, conflict, and stimulation-seeking behaviors, and the tendency to intensely value or devalue individuals. Anterior cingulate problems may also exist, evidenced by the obsessive thinking, cognitive inflexibility, and a very strong tendency to hold onto grudges and past hurts. As well, there may be temporal lobe abnormalities. The left temporal lobe is involved with aggressive behaviors toward the self and others.

Consistency and control over impulsivity are necessary to developing and sticking to character goals. When you are controlled by your emotions, constantly reacting to outside events in the heat of the moment, you cannot develop an overall sense of who you are, what you want, and how you will get it. Contemplation is important to developing a sense of right and wrong, what is good and bad for you and for others. Likewise, being enslaved by impulses and reactions denies you the opportunity to build a strong sense of self-esteem. When you can control what you do, you can feel greater certainty about your identity. It's rewarding to be able to clarify your personal values and to stick to them, to know that you and you alone are in charge of your life.

It's hard to build a sense of security and of being loved when you find yourself attaching unrealistic expectations to people to

204 • DANIEL G. AMEN

whom you're attracted and then ending friendships before they've had a chance to develop. Social connectedness takes work: It implies forgiving and flexibility. It's important for all of us to try to develop greater empathy for others by asking ourselves about another's point of view and not to automatically assume we know what others feel and think.

When Is It Time to See a Professional About My Brain?

This question is relatively easy to answer. People should seek professional help for themselves or a family member when their behaviors, feelings, thoughts, or memory (all brain functions) interfere with their ability to reach their potential in their relationships or work. If you are experiencing persistent relationship struggles (parent-child, sibling, friend, partner), it's time to get help. If you have ongoing work problems related to your memory, moods, actions, or thoughts, it is time to get professional help. If your impulsive behavior, poor choices, or anxiety are causing consistent monetary problems, it's time to get help. Many people think they cannot afford to get professional help. I think it is usually much more costly to live with brain problems than it is to get appropriate help.

Pride and denial can get in the way of seeking proper help. People want to be strong and rely on themselves, but I am constantly reminded of the strength it takes to make the decision to get help. Also, getting help should be looked at as a way to get your brain operating at its full capacity.

Angela came to see me for temper problems. Even though she was very competent at work, her behavior at home often caused problems with her husband. When her husband suggested she see me, she resisted. There was nothing wrong with her, she thought, it was everyone else. One day, after exploding at one of her children, she realized it was, at least partly, her fault and agreed to come for help. She resisted because she did not want to be seen as weak or defective. The brain-SPECT scan helped her to see that her brain needed to be balanced. With the appropriate help, she got better and didn't have to suffer from mood swings, and she and her family suffered less stress as a result of her better-balanced brain.

What to Do When a Loved One Is in Denial About Needing Help

Unfortunately, the stigma associated with a "mental illness" prevents many people from getting help. People do not want to be seen as crazy, stupid, or defective and do not seek help until they (or their loved one) can no longer tolerate the pain (at work, in their relationships, or within themselves). Most people do not see psychiatric problems as brain problems, but rather as weak character problems. Men are especially affected by denial.

Many men, when faced with obvious troubles in their marriages, their children, or even themselves, are often unable to really see problems. Their lack of awareness and strong tendency toward denial prevent them from seeking help until more damage than necessary has been done. Many men have to be threatened with divorce before they seek help. Some people may say it is unfair to pick on men. And, indeed, some men see problems long before some women. Overall, however, mothers see problems in children before fathers and are more willing to seek help, and many more wives call for marital counseling than do husbands. What is it in our society that causes men to overlook obvious problems, or to deny problems until it is too late to deal with them effectively or until more damage was done than necessary? Some of the answers may be found in how boys are raised in our society, the societal expectations we place on men, the overwhelming pace of many men's daily lives, and in the brain.

Boys most often engage in active play (sports, war games, video games, etc.) that involves little dialogue or discussion. The games often involve dominance and submissiveness, winning and losing, and little interpersonal communication. Force, strength, or skill handles problems. Girls, on the other hand, often engage in more interpersonal or communicative types of play, such as dolls and storytelling. Fathers often take their sons to throw the ball around or shoot hoops, rather than to go for a walk and talk.

Many men retain the childhood notions of competition and the idea that one must be better than others to be any good at all. To admit to a problem is to be less than other men. As a result, many men wait to seek help until their problem has become obvious to the whole world. Other men feel responsible for all that happens in their families, so admitting to a problem is the same as admitting that they have in some way failed.

Clearly, the pace of life prevents many people and particularly men from taking the time to look clearly at the important people in their lives and their relationships with them. When we spend time with fathers and husbands and help them slow down enough to see what is really important to them, more often than not they begin to see the problems and work toward helpful solutions. The issue is generally not one of being uncaring or uninterested; it is not seeing what is there. Men are wired differently than women. Men tend to be more left brained, which gives them better access to logical, detail-oriented thought patterns. Women tend to have greater access to both sides of their brains, with the right side being involved in understanding the gestalt or big picture of a situation. The right side of the brain also seems to be involved in being able to admit to a problem. Many men just don't see the problems associated with anxiety or depression even though the symptoms may be very clear to others.

Here are several suggestions to help people who are unaware of or unwilling to get the help they need. Try the straightforward approach first (but with a new brain twist). Clearly tell the person what behaviors concern you, and explain that the problems may be due to underlying brain patterns that can be easily tuned up. Tell them help may be available—not help to cure a defect but rather help to optimize how their brain functions. Tell them you know they are trying to do their best, but their behavior, thoughts, or feelings may be getting in the way of their success (at work, in relationships, or within themselves). Emphasize better function, not defect.

Give them information. Books, videos, and articles on the subjects you are concerned about can be of tremendous help. Many people come to see us due to a book, video, or article. Good information can be very persuasive, especially if it is presented in a positive, life-enhancing way.

When a person remains resistant to help, even after you have been straightforward and given them good information, plant seeds. Plant ideas about getting help and then water them regularly. Drop an idea, article, or other information about the topic from time to time. If you talk too much about getting help, people become resentful and stubbornly won't get help, especially the overfocused types. Be careful not to go overboard.

Protect your relationship with the other person. People are more receptive to people they trust than to people who nag and belittle them. Work on gaining the person's trust over the long run. It will make them more receptive to your suggestions. Do not make getting help the only thing that you talk about. Make sure you are interested in their whole lives, not just their potential medical appointments.

Give them new hope. Many people with these problems have tried to get help and it did not work or it made them even worse. Educate them on new brain technology that helps professionals be more focused and more effective in treatment efforts.

There comes a time when you have to say enough is enough. If, over time, the other person refuses to get help, and his or her behavior has a negative impact on your life, you may have to separate yourself. Staying in a toxic relationship is harmful to your health, and it often enables the other person to remain sick as well. Actually, I have seen that the threat or act of leaving motivates people to change, whether it is about drinking, drug use, or treating ADD. Threatening to leave is not the first approach I would take, but after time it may be the best approach. Realize you cannot force a person into treatment unless they are dangerous to themselves, dangerous to others, or unable to care for themselves. You can only do what you can do. Fortunately, there is a lot more we can do today than even ten years ago.

Finding a Competent Professional Who Uses This New Brain Science Thinking

The Amen Clinics get many calls, faxes, and e-mails each week from people all over the world looking for competent professionals who think in similar ways to the principles outlined in this book. Because this approach is on the edge of what is new in brain science, other professionals who know and practice this information may be hard to find. However, finding the right professional for evaluation and treatment is critical to the healing process. The right professional can have a very positive impact on your life. The wrong professional can make things worse.

There are a number of steps to take in finding the best person to assist you. The right help is not only cost effective but saves unnecessary pain and suffering, so don't rely on a person simply because they are on your managed care plan. That person may or may not be a good fit for you. Search for the best. If he or she is on your insurance plan, great, but don't let that be the primary criteria. Once you get the names of competent professionals, check their credentials. Very few patients ever check a professional's background. Board certification is a positive credential. To become board certified, physicians must pass additional written and verbal tests. They have had to discipline themselves to gain the skill and knowledge that was acceptable to their colleagues. Don't give too much weight to the medical school or graduate school the professional attended. I have worked with some doctors who went to Yale and Harvard who did not have a clue on how to appropriately treat patients, while other doctors from less prestigious schools were outstanding, forward thinking, and caring. Set up an interview with the professional to see whether or not you want to work with him or her. Generally you have to pay for their time, but it is worth spending the money to get to know the people you will rely on for help.

Many professionals write articles or books or speak at meetings or local groups. Read the work of or hear the professional speak, if possible. By doing so, you may be able to get a feel for the person and his or her ability to help you. Look for a person who is open-minded, up-to-date, and willing to try new things. Look for a person who treats you with respect, who listens to your questions, and responds to your needs. Look for a relationship that is collaborative and respectful. I know it is hard to find a professional who meets all of these criteria who also has the right training in brain physiology, but these people can be found. Be persistent. The caregiver is essential to healing.

Do not let pride get in the way of getting the help you need. In order to make a good brain great, you have to admit when you need help.

Lesson #11: Fix the issues that get in the way of great sex.

A HEALTHIER BRAIN
EQUALS A SEXIER YOU

—

Strategies to Improve Your Brain and Life

You know you've got to exercise your brain just like your muscles.

—WILL ROGERS

I am a sucker for a beautiful brain. Usually one needs to be attracted physically to a potential romantic partner, but my brain-imaging work has taught me that it is also a good idea to be attracted to the appearance of a person's physical brain as well. Ugly brains usually make for ugly relationships. "Beauty and brains" is more than just a cliché. In 2001, CNN International aired a story on my imaging work. News anchor Marina Kolbe spent several days in our clinic watching us work and filming the imaging process; she was even scanned herself as part of our healthy-brain study. Besides being a smart, attractive woman, she also had one of the prettiest brains I had ever seen. Now, to a neuroscientist, that is ever so sexy. She and I have been friends ever since. Her behavior is consistent with her lovely brain. Beauty is much more than skin deep.

Roseanne has been a friend of mine for many years. She is an attractive woman, but not attractive to me. I saw her as anxious,

THE BRAIN IN LOVE · 211

worried, and fretful. One day her doctor put her on the antidepressant Zoloft to calm her anxiety. Several weeks later I found myself being more interested in her. Something about her was different. Even though she looked the same, she was more appealing. She had an air of confidence that had been missing before the medication. Her smile was brighter and she seemed to have a more genuinely positive internal state. Her eyes had a new, more intense sparkle. Her brain was more relaxed, one of the effects of Zoloft. A healthier brain is associated with a healthier, sexier you.

Since your brain is involved in everything you do, including everything sexual, it follows that a healthy brain is more likely to be associated with more effective behavior, at work, at home, and even in the bedroom. A healthy brain will give you more consistently loving behavior, help you read social cues, and allow you to be a better lover. As in Roseanne's case, it can even increase your sex appeal. Working to keep your brain healthy increases your chances for loving relationships and great sex.

In my work I have seen many things that hurt how the brain functions, ruining your chances for love; and many things that help brain function, improving your chances for love. In this chapter I will explore behaviors that make your brain look old, ugly, shriveled, and damaged as well as those things that enhance and beautify the brain. Once you know what helps and hurts the brain, you will have a clearer choice on how healthy you want your brain to be, and subsequently how effective you will be.

Hurtful Brain Behaviors

Brain Trauma

One of the most important lessons I have learned from looking at 35,000 scans these last sixteen years is that mild traumatic brain injuries change people's whole lives and virtually nobody knows about it. Virtually no marital counselor on the planet thinks about evaluating brain trauma as part of why people struggle in their relationships. Even what many professionals would consider mild trauma can be harmful. The brain is the consistency of soft butter or tofu, somewhere between egg whites and Jell-O. It is housed in a really hard skull that has many ridges.

I recently scanned a nineteen-year-old man who had a skateboarding accident which caused him to be unconscious for about half an hour. Most physicians would consider that a mild traumatic brain injury. Yet on his scan it literally wiped out 25 percent of the front part of his brain; so the part of his brain that is involved with judgment, forethought, impulse control, organization, and planning is dead, from what professionals consider a minor brain injury. Protecting the brains of our children, our loved ones, and ourselves should be a top priority because brain damage affects us in a very serious way. A high percentage of people in prison have had brain injuries. There is a higher incident of children who struggle in school after a brain injury. There is a higher incident of depression and substance abuse after brain injuries.

I treat a couple in which the woman had a diving accident when she was eighteen years old, again something that was considered a mild traumatic brain injury as she was unconscious for only a short period of time. She was dating her soon-to-be husband at the time, and he noticed a big change in her behavior over the next six months. She went from being loving, attentive, sweet, consistent, and reliable to someone who was more emotional, depressed, disorganized, and temperamental. He felt committed to her and hoped she would change back to the person she used to be. Unfortunately, she struggled with her mood and her temper for thirty years, until she came to see me. When I scanned her, she was missing the function of about 30 percent of her left prefrontal cortex, which is the happy side of the brain. The left side of the brain has been reported by researchers to be the happy side, while the right side is associated with more anxiety and negativity. When you hurt the happy side of the brain, the more anxious, negative right side has more dominance in your life. This injury seriously affected this couple's happiness.

I have discovered that you have to ask people five to ten times whether or not they have had brain trauma. People just forget, even serious injuries. I have a friend who has financial problems that have affected his marriage. His ability to have sex suffers due to his financial problems. His wife is anxious about money and she feels insecure about their future. He is an amazing, sweet man, but struggles to make good financial decisions. Before I scanned him, I asked him seven or eight times whether or not he had had a head injury, which is common in people who struggle with finances. He said no, no, no, no, no, no. When I scanned him, I saw a large dent in the function of the front part of his brain. The only thing that I could figure is that he had a head injury that he forgot. So I asked him again and he said no. I asked him again and was very specific, "Have you ever been in a car accident?"

"No," he said.

And then, and it always happens this way, his faced changed, he got this "Aha" look on his face, and said, "I am so sorry. I lied to you. When I was fifteen years old I was in the front passenger seat of a car when we got in a head-on collision. I wasn't wearing my seat belt and my head broke the windshield. I lost my eyesight for four days. I can't believe I forgot that incident." He hurt a significant portion of the judgment center in his brain. He's still a wonderful, sweet, loving man, but when it comes to money, he would make bad decisions that increased his wife's anxiety and decreased her libido.

Having a head injury can really ruin your chances for great sex. Protect your brain. I work with many couples who have experienced domestic violence. The incidence of head injuries in violent individuals is rampant and very few people know it. If you have serious problems with your temper and the police come to your house, the first thing the courts and mental health professionals recommend is anger-management classes. If a head injury is part of the cause of the trouble, anger-management classes are like trying software programs to fix hardware problems. Not very effective.

I once treated a man from Normal, Illinois, who was referred to

214 · DANIEL G. AMEN

me by his psychologist after I gave a lecture at the university there. What a fun visit for this California boy. I went to the Normal grocery store, was interviewed on the Normal radio station, and even had the opportunity to meet Normal women—not many normal women in California. . . . The patient had been arrested for felony domestic violence. He had broken his wife's arm. When I met this man, he truly hated himself. He hated his temper and his inability to control the rage he felt inside. He was suicidal at our first visit. His SPECT scan showed a dent in the left-front side of his brain and he had very poor function in the part of his brain called the left temporal lobe, an area that we have associated with violence. I had already asked him six times if he had had a head injury, to which he replied no, but when I had evidence on the scan of a head injury and asked if he had ever fallen out of a tree, fallen off a fence, or dived into a shallow pool, an "Aha" look came over his face and he said, "When I was six years old I was standing on top of the railing on our porch, it was raining, and I slipped and fell six feet head first into a pile of bricks. My parents told me I was only unconscious for a little bit. Do you think that could cause this problem?" I said it could and asked when the temper problems began. He said he had always had them. I then asked him to ask his mom if he had them before he was in kindergarten. It turned out that his temper problems started when he was in first grade. He had problems nearly his whole life, probably secondary to a brain injury nobody knew about. He and his wife had been to multiple relationship counselors with no benefit. If you never look at the brain, you may miss very important pieces of information. The combination of the right medication changed his life. The other interesting thing is that it changed his wife's life as well because rather than seeing him as a bad person, she saw him as somebody who was sick, that potentially with the right treatment could get much better.

Emotional Trauma

In a similar way, emotional trauma can change the brain negatively and make it harder for you to get the love that you want.

Whenever people have been physically, emotionally, or sexually abused, brain changes take place. Being in a fire, car accident, earthquake, or flood can also change the brain. In our imaging work we have seen specific scan patterns associated with emotional trauma. The limbic or emotional centers of the brain tend to become overactive, making people vulnerable to obsessions, anxiety, and depression, all things that interfere with our ability to connect with others in a loving way. It has been well documented that adverse childhood events affect people throughout their life. These experiences disrupt the child's ability to form secure attachments to their parents. This may lead to their inability to form secure attachments later in life. These people have multiple partners and are promiscuous in their sexuality.

As with brain injuries, many people forget they have had emotional trauma. When we see the scan pattern associated with trauma, we ask about it multiple times. I have been surprised many times by people who said they never had trauma, later to remember being sexually molested, in a fire, robbed at knife point, raped, or even attacked by animals.

One of our patients, who saw a colleague of mine at the clinic, had the scan pattern of emotional trauma. She initially denied any past trauma. When given more specific examples, such as being in a fire, robbed, or raped, she said no, no, no. Then after a long hesitation she said that there was this one time when she was ten years old and went to a friend's house in the high desert of California. Her friend's father was an actor who collected unusual animals. They had a lion at home and the day she was at her friend's house, the lion got loose and chased her, pinned her, and actually had her head in his mouth before they got him off of her.

Emotional trauma can impact the brain and wreak havoc in your love life. Getting the emotional trauma treated can actually rebalance the brain and improve your chances for love. Many researchers have seen that early abuse survivors have overlapping psychiatric disorders. Therefore, they may suffer from major depressive disorder, generalized anxiety disorder, and panic disorder. Emotional trauma, like abuse or exposure to domestic violence, at early ages has been studied extensively and has shown that negative behaviors have resulted later in life. Depression, suicide, and drug abuse in later life are often associated with trauma early in life.

Drug and Alcohol Abuse

Drugs and alcohol clearly damage the brain. They prematurely age and lower overall function in the brain. Alcohol is toxic to the brain if you drink more than a couple of drinks a week. A study from Johns Hopkins University reported that people who drink every day have smaller brains. When it comes to the brain, *size really does matter*. Alcohol kills cells in the cerebellum, the back bottom part of the brain that is involved with coordination, learning, and orgasmic pleasure.

Many people drink and use drugs as a way to medicate negative feelings. If you have experienced emotional trauma, you are more likely to drink. In fact, up to 30 percent of people with alcohol abuse or drug abuse have emotional trauma in their backgrounds. As many as 60 percent of women who are substance abusers have posttraumatic stress disorder (PTSD), which is associated with nightmares, flashbacks, emotional numbing, anxiety, insomnia. When they stop their drug or alcohol abuse, their PTSD often gets worse. People also medicate social anxiety with marijuana, painkillers, or alcohol. They feel better in social situations, less inhibited. The problem with using substances for self-medication is that they actually damage brain function and subsequently damage the parts of the brain involved in forethought, judgment, impulse control, organization, and planning—all things important for healthy sexual behavior.

I once saw a woman from Maine who had problems with obsession and anxiety. Her husband came along and got scanned, in his mind, just to support his wife. I looked at her brain and saw the trouble we expected and prescribed a course of treatment. When I looked at his fifty-six-year-old brain, his brain looked like he was eighty. I asked him what he was doing to hurt his brain.

"Nothing, Dr. Amen," he said.

I said, "Really? How much do you drink?"

"Oh, not very much," he replied.

"What's not very much?"

"Oh, maybe I have three or four drinks a day."

"Every day?" I said.

"Yeah, every day. But it's never a problem. I never get drunk. I have never gotten into trouble with it," he said with anxiety.

I said, "Why do you drink every day?"

"Since my son went off to college, I have this empty-nest thing going on. I just get great enjoyment out of going to the bar, seeing my friends. It's a social time, kind of like the show *Cheers*."

I said, "Well, you are poisoning yourself. You're fifty-six and you're brain looks like it's eighty. If you keep this up, pretty soon a lot of your brain is going to be dead."

It shocked him that his brain looked as bad as it did. Then he developed this concept I call brain envy. After learning about the brain, he wanted a better one. I helped him develop a brain-healthy plan that included abstinence from alcohol, regular exercise, mental exercise, vitamins, and fish oil. Four months later he wrote me back saying that he mentally felt like he was twenty. His energy and memory were better, he felt smarter, more articulate. His work as a writer had also improved.

Lisa, forty-two, drank three to four glasses of wine nearly every day. She rarely got drunk, but felt uncomfortable when she didn't drink. Her husband noticed over the past few years that she was not herself. She was more forgetful and more irritable. She started to have high blood pressure and was much less sexual and less sexually responsive. I saw her out of concern for her memory problems. Her SPECT scan showed overall decreased activity. She had a toxic brain. The alcohol was damaging her brain, affecting her memory and moods, and even her sexuality. High blood pressure is a common side effect of too much alcohol. With high blood pressure, blood flow to vital organs, such as the brain and genitals, is impaired, so there will be trouble having an orgasm and trouble thinking, a bad combination.

Sometimes small amounts of alcohol can calm anxiety and help people be more receptive to social and sexual interactions. Small amounts of alcohol have been found useful for a number of issues, including heart health. But small amounts mean one or two glasses a week, not a day as most people think. A little bit of alcohol is not my concern in this section.

My favorite definition of an alcoholic or drug addict is anyone who has gotten into trouble (legal, relational, or work related) while drinking or using drugs, then continues to use them. They did not learn from the previous experience. A rational person would realize that he or she has trouble handling the alcohol or drugs and stay away from them. Unfortunately, many people with these problems have to experience repeated failures because of the substance use, and thus hit "rock bottom" before treatment is sought.

In treating substance abuse, it is important to recognize and treat any underlying cause of the problem, such as unrecognized depression, bipolar disorder, anxiety disorders, or ADD. New medications have been developed that have been found helpful in alleviating withdrawal symptoms and decreasing cravings for the substances. Psychotherapy and support groups are often helpful.

Drugs and alcohol are a nightmare for brain function.

Toxic Exposure

Toxic exposure hurts the brain. Many substances have the potential to be brain toxic and most people have no clue. Many medications, much caffeine, nicotine, and environmental toxins, such as pesticides, paint fumes, hair-coloring chemicals, and nail polish, can hurt your brain. Understanding the sources of brain poisons can help you avoid them.

Many medications are brain toxic. From a psychiatric standpoint, I was taught to use a class of antianxiety medications called benzodiazepines, such as Xanax, Ativan, and Valium, to treat patients with intense feelings of anxiety and panic. As soon as I started performing SPECT studies, I saw that these medications

were often toxic to brain function. Scan after scan on these medications showed an overall diminished or dehydrated pattern of
activity, just as there is with drug abuse. It didn't take long for me
to stop using these medications and look for other ways to heal
anxiety and panic. In much the same way, painkillers often showed
brain toxicity on scans: Vicodin, Darvon, Percodan, OxyContin,
and others caused overall decreased brain activity. No wonder they
help pain, they make people feel numb all over. Of course, this
doesn't mean that these medications are never indicated. Many
people would rather die than live with chronic pain. It does mean,
however, that we should look for alternatives to painkillers for
chronic pain that also numb the brain.

Nicotine prematurely ages the brain. Nicotine, found in ciga-
rettes, cigars, chewing tobacco, and nicotine patches, tablets, and
gum, causes blood vessels to lessen blood flow to vital organs. We
know smokers have more problems with impotence; it is bad to
have low blood flow to sexual organs. Nicotine constricts blood
flow to the skin, making smokers look prematurely older than they
are. Nicotine also constricts blood flow to the brain, eventually
causing overall lowered activity and depriving the brain of the
nutrients it needs.

High amounts of caffeine can also be trouble. Caffeine constricts blood flow to the brain and many other organs. A little caffeine a day is not a problem, but more than a cup or two can be
trouble. Caffeine does three bad things to the brain. First, brain-
imaging studies have shown that caffeine constricts blood flow to
the brain. Since blood is critical in bringing nutrients to cells and
taking away toxic waste products, anything that diminishes blood
flow to an organ causes premature aging. Second, caffeine blocks
a chemical called adenosine and fools us into believing we need
less sleep. Less sleep also causes overall lower blood flow to the
brain. In addition, caffeine is a diuretic. The brain is 80 percent
water. Anything that dehydrates you has a negative impact on
brain function. As little caffeine as possible is a good rule if you
want to respect and nurture your brain.

Sleep Deprivation

Sleep deprivation hurts the brain. People who get less than seven hours of sleep a night have lower activity in the temporal lobes of the brain, the part of the brain involved in learning and memory. Shift workers, those suffering from jet lag, teens who have their sleep schedules off-kilter from school schedules, and those suffering from sleep apnea are all at risk for poorer brain function. Those who are sleep deprived score poorer on memory and math tests, have lower grades in school, and are at much greater risks for driving accidents. Sleep deprivation is also associated with depression and attention deficit disorders. Recently, sleep apnea (snoring loudly, holding breath when sleeping, and tiredness during the day) has been linked to Alzheimer's disease. This is due to the direct effect of a lack of oxygen damaging the brain. Work to sleep at least seven to eight hours a night. Practice good sleep habits, such as avoiding much caffeine or nicotine; staying away from alcohol as a sleep aide, as it will wear off and cause you to wake up in the middle of the night; and avoiding exercise before bed; and learn relaxation techniques to calm your mind.

Helping your partner get enough sleep is an effective form of foreplay. Helping with the children, work, house chores, or other things that interfere with sleep is likely to help your partner sleep better and be more available to you.

Untreated Mental Illnesses

As in the case of Roseanne earlier in the chapter, being plagued with depression, anxiety, distractibility, impulsivity, mood swings, or obsessions is not very sexy. Having an untreated psychiatric problem can ruin your chances for great love. Many people never seek help for their mental or emotional problems because they do not recognize them or think they can control them. All aspects of sexuality may be affected by the distress of mental illness. In both men and women, people who were depressed had half the sexual arousal than their normal counterparts did. Mental illness is very common, almost the norm. According to the U.S. Epidemiological Catchment Area Study, 49 percent of the U.S. population will experience a mental illness at some point in their lives, with anxiety, depression, attention deficit disorders, and substance abuse being the most common. If you suffer from emotional or behavior problems, get the help.

Poor Diets

The increase in fast-food diets and poor nutrition is directly responsible for the rise in mental illness over the past fifty years. Studies have linked attention deficit disorder, depression, Alzheimer's disease, and schizophrenia to junk food and the absence of essential fats, vitamins, and minerals in industrialized diets. These illnesses have a direct impact on sexual and relationship health. Rates of depression have been shown to be higher in countries with low intakes of fish, for example. Lack of folic acid, omega-3 fatty acids, selenium, and the amino acid tryptophan are thought to play an important role in the illness. Deficiencies of essential fats and antioxidant vitamins are also thought to be a contributory factor in schizophrenia.

Helpful Brain Behaviors

Protect Your Brain

Protecting the brain from injury is the first step to optimizing its function. Your brain is the consistency of tofu and it is housed in a really hard skull. Wear a seat belt in a car and always wear a helmet when you ride a bicycle, motorcycle, or go snowboarding.

Limit Brain Toxins

As mentioned above, current research has shown that many chemicals are toxic to brain function. Alcohol, drugs of abuse, nicotine, much caffeine, and many medications decrease blood flow to the brain. When blood flow is decreased, the brain cannot work efficiently. Avoid these toxic substances.

Get Enough Sleep

Sleep deprivation decreases brain activity and limits access to learning, memory, and concentration. A recent brain-imaging study showed that people who consistently slept less than seven hours had overall less brain activity. Getting enough sleep is essential to brain function.

Counteract Stress

Scientists have only recently discovered how stress negatively affects brain function. Stress hormones have been shown in animals to be directly toxic to memory centers. Brain cells can die with prolonged stress. Managing stress effectively through meditation, relaxation, and exercise is essential to good brain function.

Eat Right to Think Right

The fuel you feed the brain has a profound effect on how it functions. Lean protein, complex carbohydrates, and foods rich in omega-3 fatty acids (large cold-water fish, such as tuna and salmon, walnuts, Brazil nuts, olive oil, and canola oil) are essential to brain function. Research has shown that fish consumption is associated with the prevention of cognitive decline as people age.

Daily multiple vitamins. With our poor diet, many Americans are overweight but nutrient poor. The American Medical Association recommends we all take a 100 percent multiple vitamin everyday. Your brain needs it.

Daily high-quality fish oil. Omega-3 fatty acids are essential to brain health. Low levels of this nutrient have been associated with depression, ADD, dementia, and even suicide. Fish oil has been found to be helpful for hearts, joints, skin, and brain. Take 1,000 to 2,000 milligrams a day.

Kill the ANTs That Invade Your Brain

The quality of your thoughts also impacts brain function. Happy, hopeful, positive thoughts are associated with improved brain function, while negativity (I call these bad thoughts ANTs, automatic negative thoughts) turns off certain cerebral centers. Positive thinking is not just good for you, it helps make your brain work better. List five things you are grateful for today.

Work Your Brain

Your brain is like a muscle. The more you use it, the more you can use it. Every time you learn something new, your brain makes a new connection. Learning enhances blood flow and activity in the brain. If you go for long periods without learning something new, you start to lose some of the connections in the brain and you begin to struggle more with memory and learning. Strive to learn something new every day, even if it is just for a short period of time.

Exercise for Your Brain

People who physically exercise on a regular basis have better memories with age, they have better blood flow to the brain, and many cerebral processes are enhanced. The best kind of exercise improves the pump force of your heart (cardiovascular exercise).

It has been known for many years that sex was good exercise, but until now nobody had made a scientific study of the caloric content of different sexual activities. An anonymous source on the Internet listed, tongue in cheek, the following results.

REMOVING HER CLOTHES

With her consent	12 calories
Without her consent	2,187 calories

OPENING HER BRA

With both hands	8 calories
With one hand	12 calories
With your teeth	485 calories

PUTTING ON A CONDOM

With an erection	6 calories
Without an erection	3,315 calories

POSITIONS

Missionary	12 calories
69 lying down	78 calories
69 standing up	812 calories
Wheelbarrow	216 calories
Doggy style	326 calories
Italian chandelier	2,912 calories

ORGASMS

Real	112 calories
Fake	1,315 calories

POSTORGASM

Lying in bed hugging	18 calories
Getting up immediately	36 calories
Explaining why you got out of bed immediately	816 calories

GETTING A SECOND ERECTION IF YOU ARE . . .

20–29 years	36 calories
30–39 years	80 calories
40–49 years	124 calories

50–59 years	1,972 calories
60–69 years	7,916 calories
70 and over	Results are still pending

DRESSING AFTERWARD

Calmly	32 calories
In a hurry	98 calories
With her father knocking at the door	5,218 calories
With your wife knocking at the door	13,521 calories

Results may vary!

Develop a "concert state" for your brain

Optimal performance is best achieved when a "concert state" exists in the brain. By "concert state" I mean "a relaxed body with a sharp, clear mind," much as you would experience at an exhilarating symphony. Achieving this state requires the ability to relax and focus. Ten minutes of slow, deep breathing is usually enough to develop this state.

Coordinate Your Brain

Recently, it was learned that the cerebellum, a structure at the back, bottom part of the brain, is involved not only with physical coordination but also with processing speed and thought coordination, or how quickly you can integrate new thoughts. Doing coordination exercises helps the brain overall work smarter and faster. One exercise shown in scientific studies to help is juggling. People who learned how to juggle three balls for a minute straight enhanced brain function.

Make Love for Your Brain

As discussed in Lesson Two, making love on a regular basis improved mood, memory, and overall health. One study showed that it decreased the risk of heart attack and stroke by 50 percent. Hold the medicine, give me love.

Do Not Let E-mail Control Your Life

A recent study showed that constantly checking e-mails and phone messages actually lowered IQ by ten points, more than double the loss in IQ of cannabis users. E-mail can be addictive, as one is always waiting for the next good e-mail to hit, like waiting for the next blackjack in the card game twenty-one. The anticipation of something good keeps us checking something routine. It also distracts us from staying focused on the person or task at hand. Limit your time on e-mail.

Treat Brain Problems Early

If you have learning, mood, behavior, or memory problems, get an evaluation. The brain is an organ just like your heart or kidneys. When there are signs of trouble, work on getting it fixed.

Use these strategies to have the best brain possible. A better brain improves everything else in your life, including your sex life.

Lesson #12: Caring for your brain increases your chances for great sex.

The Amen Clinic Brain System Questionnaire

Please rate yourself on each of the symptoms listed below using the following scale. If possible, for the most complete picture, have another person who knows you well (such as a spouse, lover, or parent) rate you also.

0	1	2	3	4
Never	Rarely	Occasionally	Frequently	Very Frequently

PREFRONTAL CORTEX ISSUES (PFC)

1. Fails to give close attention to details or makes careless mistakes
2. Has trouble sustaining attention or trouble listening
3. Is poorly organized
4. Is easily distracted
5. Finds difficulty expressing empathy for others
6. Blurts out answers before questions have been completed, interrupts frequently
7. Is impulsive (saying or doing things without thinking first)

ANTERIOR CINGULATE GYRUS (ACG)

8. Excessive or senseless worrying
9. Upset when things do not go your way
10. Upset when things are out of place
11. Tendency to be oppositional or argumentative
12. Tendency toward compulsive behaviors
13. Intense dislike for change
14. Needing to have things done a certain way or you become very upset

DEEP LIMBIC SYSTEM (DLS)

15. Frequent feelings of sadness or moodiness
16. Negativity or irritability
17. Low energy
18. Decreased interest in things that are usually fun or pleasurable
19. Feelings of hopelessness, helplessness, worthlessness, or guilt
20. Crying spells
21. Chronic low self-esteem

BASAL GANGLIA (BG)

22. Frequent feelings of nervousness, anxiety, or panic
23. Symptoms of heightened muscle tension (headaches, sore muscles, hand tremor)
24. Tendency to predict the worst
25. Conflict avoidance
26. Excessive fear of being judged or scrutinized by others
27. Excessive motivation
28. Lack of confidence in abilities

TEMPORAL LOBES (TLs)

29. Short fuse or periods of extreme irritability
30. Frequent misinterpretation of comments as negative when they are not

31. Frequent periods of déjà vu (feelings of being somewhere you have never been)
32. Sensitivity or mild paranoia
33. History of a head injury or family history of violence or explosiveness
34. Dark thoughts, may involve suicidal or homicidal thoughts
35. Periods of forgetfulness or memory problems

ANSWER KEY

Prefrontal cortex (PFC) symptoms, 1–7: _____
Anterior cingulate gyrus (ACG) symptoms, 8–14: _____
Deep limbic system (DLS) symptoms, 15–21: _____
Basal ganglia (BG) symptoms, 22–28: _____
Temporal lobe (TL) symptoms, 29–35: _____

In each system, the following number of questions with the answer of 3 or 4 indicates that problems may be present in that area.

5 questions = Highly probable
3 questions = Probable
2 questions = May be possible

See Lesson Two for a description of the brain systems, problems, and treatments.

Why SPECT?

―――

What Brain-SPECT Imaging Can Tell Clinicians and Patients That Cannot Be Obtained Elsewhere

If we agree that mental disorders and difficult behaviors may be related to functional problems in the brain, then a logical next step is clearly to consider physically evaluating the *brain itself* when faced with people who struggle with complex problems or who are unresponsive to our best diagnostic and treatment efforts. Why are psychiatrists the only physicians who rarely look at the organ they treat?

It is time to change. Amen Clinics, Inc. (ACI) has provided leadership and understanding on the clinical use of brain imaging in psychiatry. Since 1991, ACI has built the world's largest database of brain scans related to emotional, learning, and behavioral problems. The study we do is called brain-SPECT imaging. SPECT stands for Single Photon Emission Computed Tomography. It is a nuclear medicine procedure widely used in medicine to study heart, liver, thyroid, bone, and brain problems. Brain-SPECT imaging is a proven, reliable measure of cerebral blood flow. Because brain activity is directly related to blood flow, SPECT effectively shows us the patterns of activity in the brain.[1] SPECT allows physicians to

―――

1. Holman, B. L., and M. D. Devous Sr. "Functional Brain SPECT: The Emergence of a Powerful Clinical Method." *Journal of Nuclear Medicine* 1992; 33:1888–1904.

look deep inside the brain to observe three things: areas of the brain that work well, areas of the brain that work too hard, and areas of the brain that do not work hard enough. ACI has performed more than 35,000 scans on patients from age 10 months to 101 years, and has also scanned many normal, "healthy brain" individuals as well.

The procedure guidelines of the Society of Nuclear Medicine lists the evaluation of suspected brain trauma, evaluation of patients with suspected dementia, presurgical location of seizures, and the detection and evaluation of cerebral vascular disease as common indications for brain SPECT.[2] The guidelines also say that many additional indications appear promising. At ACI, because of our experience, we have added the indications of evaluating violence, substance abuse, the subtypes of ADD, anxiety, and depression, and complex or resistant psychiatric problems for brain SPECT.

An important question for today's mental health clinicians is "When and why would I order a SPECT study for my patients or get one for myself or a loved one?" My purpose in this section is to answer this question and to point out some of the benefits and caveats for using this powerful tool. A SPECT scan can provide distinct benefits to clinicians, and to the patient and his/her family. (There are also some things that should *not* be expected from a SPECT scan, listed later.)

Benefits for Physicians and Clinicians

1. A SPECT scan can show:
   a. Areas of the brain implicated in specific problems, such as the prefrontal cortex with executive function and the medial temporal lobes with long-term memory storage.
   b. Unexpected findings that may be contributing to the pre-

2. Amen, D. *Comprehensive Textbook of Psychiatry*, edited by Kaplan and Sadock, Neuroimaging Chapter, 2000.

senting problem(s), such as toxicity, potential areas of seizure activity, or past brain trauma.

c. Potential seizure activity, in many cases more accurately seen by SPECT than standard EEG, especially in the areas of the medial temporal lobe. There are more than forty-one studies with more than 1,300 patients on SPECT and epilepsy (see www.amenclinic.com for references).

d. Targeted areas for treatment, such as overactive basal ganglia or anterior cingulate gyrus (seen on anxiety and OCD spectrum disorders) or an underactive temporal lobe (seen in seizure disorders and trauma).

e. Specific effects of medication on the brain to help guide us in adjusting dosages or augmenting treatment. Often patients report that SSRIs are helpful but also cause decreased motivation or memory problems, seen as decreased prefrontal or temporal lobe activity on SPECT.

f. Changes in brain function with treatment, improved or worsened. You can review many "before and after" scans at www.amenclinic.com.

2. The image occurs at the time of injection and outside the imaging camera, which gives SPECT several significant advantages. Most notably, we are able to sedate people after they have been injected so that they can lie still for the scan, often difficult for hyperactive or autistic children or demented adults (motion artifact ruins the scan in all of these imaging techniques).

3. A SPECT scan can provide explanations for refractory symptoms and help clinicians ask better and more targeted questions (e.g., about toxic exposure, brain injuries, anoxia, inflammation, or infections that patients may have denied or forgotten).

4. A SPECT scan can help us to avoid prescribing treatments that make the problem worse, such as unnecessarily stimulating an already overactive brain or calming an underactive one.

5. A SPECT scan can help to evaluate risk for dementia—the brain starts to change long before people show symptoms. There is usually a loss of 30 percent of hippocampal tissue before symptoms occur. Using autopsy data in fifty-four patients, Bonte reported that brain SPECT had a positive predictive value for Alzheimer's disease of 92 percent.[3]

6. A SPECT scan can also help to differentiate between types of dementia. Early in the disease, Alzheimer's disease, frontal temporal lobe dementia, Lewy body dementia, and multiinfarct dementia each have their own patterns. There are more than eighty-three studies with more than 4,500 patients on this subject (see www.amenclinic.com for references).

7. A SPECT scan helps clinicians understand the rationale for using certain medications (such as anticonvulsants to stabilize temporal lobe function or to calm focal areas of marked hyperactivity, or stimulants to enhance decreased prefrontal perfusion, or SSRIs to calm basal ganglia and anterior cingulate hyperactivity).

8. A SPECT scan can identify specific areas of the brain affected by trauma, better target treatment, and help deal with insurance, legal, and rehabilitation issues. There are more than thirty-eight studies with more than 1,300 patients on brain trauma (see www.amenclinic.com for references).

9. A SPECT scan can often identify factors contributing to relapse in recovering substance abusers, eating disordered, or sexual addicts. For example, the patient may have suffered an injury to the prefrontal cortex or temporal lobes or have overactivity in the anterior cingulate gyrus, basal ganglia, limbic system, or prefrontal cortex, each of which could indicate comorbid disorders requiring treatment.

10. A SPECT scan can often identify a specific cause or reason that contributes to recovering alcoholics, drug addicts, eat-

3. Bonte F. J., M. F. Weiner, E. H. Bigio, and C. L. White. "Brain Blood Flow in the Dementias." SPECT with Histopathologic Correlation in 54 Patients." *Radiology* 1997; 202:793–797.

ing disordered, or sexual addicts relapse behavior in their recovery from an addictive process. For example, the patient may have suffered an injury in the prefrontal cortex or temporal lobes or have overactivity in anterior cingulate gyrus, basal ganglia, limbic system, or prefrontal cortex, each of which could contribute to the relapsing behaviors.

11. A SPECT scan is also useful to determine if further adjustment of medication is needed. Scans of patients on medication will reveal areas of the brain still overactive or underactive.

Benefits of SPECT-Brain Imaging for Patients and Their Families

1. A SPECT scan helps develop a deeper understanding of the problem, resulting in reduced shame, guilt, stigma, and self-loathing. This can promote self-forgiveness, often the first step in healing. Patients can see that their problems are, at least in part, medical and physical.

2. A SPECT scan allows patients to see a physical representation of their problems that is accurate and reliable, and helps to increase compliance—pictures are powerful. It can influence a patient's willingness and ability to accept and adhere to the treatment program. They can better understand that not taking medication for anxiety, depression, rage, ADD, etc. is similar to not wearing the "correct" prescription glasses.

3. A SPECT scan helps families understand when things, such as permanent brain damage from an injury, will not get better, so that they can better accept the condition and provide accordingly.

4. A SPECT scan shows substance abusers the damage they have done to their own brain, thus helping to decrease denial, provide motivation for treatment, and support perseverance in sobriety.

5. A SPECT scan shows patients how treatments have impacted (improved or worsened) brain function.

6. A SPECT scan helps motivate abusive spouses to follow medication protocols by showing that there is a physical abnormality contributing to their problems.

7. A SPECT scan is useful for cancer patients suffering with a "chemotherapy toxic brain." It gives them insight into their cognitive struggles and also helps their doctors see the neurophysiologic and emotional effects of having cancer and its treatment.

8. A SPECT scan can help take modern psychopharmacology from mystery and unknown consequences to reality and more predictable outcomes.

9. A SPECT scan allows patients to understand why specific treatments are indicated, which medications are likely to be most helpful, and what other interventions may be indicated.

What a SPECT Scan Cannot Provide

Despite the many benefits that might be derived from a SPECT scan, there are clearly some things that it cannot provide. For example, a SPECT scan cannot:

1. give a diagnosis in the absence of clinical information

2. give the date of a head injury, infection, or toxic exposure

3. assess or evaluate IQ

4. assess or evaluate the guilt, innocence, motivation, or sanity of a criminal defendant

5. guarantee a perfect diagnosis, or a cure.

How SPECT Differs from MRI

A SPECT scan is similar to an MRI study in that both can show three-dimensional images and "slices" of the brain. However, whereas MRI shows the *physical anatomy* of the brain, SPECT shows brain *functional activity.* That is, SPECT yields images showing where the brain is functioning well, where it is working too hard, and where it is not working hard enough. A newer version of MRI, functional MRI or "fMRI," is also capable of showing brain activity and is used extensively in scientific research on brain function. fMRI shows instantaneous neural activity so you can see, for example, how the brain responds to a specific stimulus event. With SPECT we see brain activity averaged over a few minutes so it is better at showing the brain doing everyday activities such as concentrating, meditating, reading, etc. PET, another nuclear-imaging technique, is very similar to SPECT but is a slower and more costly imaging technique.

Ensuring High-Quality SPECT Images

Although a SPECT scan is simple from the patient's perspective, it takes considerable skill and experience to dependably generate accurate brain-SPECT images suitable for psychiatric applications. Equally important is the need for total consistency in imaging techniques among patients so that results are quantifiable, repeatable, and consistent.

Here are some of the factors that need to be considered in SPECT scans.

Variability-of-Technique Issues

Processing protocols need to be standardized and optimized. Motion can ruin a scan, so it is important that there be *no motion* on the scan. The professionals need to know how to identify and deal with image artifacts and other sophisticated technical issues.

Variability of cameras. Multiheaded cameras are clearly superior, as they can scan much faster. It takes an hour to do a scan on a single-headed camera, thirty minutes on a dual-headed camera, and fifteen minutes on a triple-headed camera.

Experience of readers. At the Amen Clinics, we have developed a standardized reading technique for which we have documented high inter- and intrarater reliability.

Image display. Scans must be clear, understandable, easily illustrative of brain function, and available to the patient on a timely basis. We believe our 3-D rendering software makes the scans easy for professionals, patients, and families to understand.

Drugs. Scans can be affected by a number of substances that need to be controlled, including medications, street drugs, and caffeine.

All of the above issues have been addressed at the Amen Clinics by carefully standardized procedures for all our SPECT scans.

Common Concerns

CONCERN: Low resolution—it is commonly said that a SPECT scan is a "poor man's PET study."

RESPONSE: With multiheaded cameras, SPECT has the same resolution as PET with considerably lower cost, better insurance coverage, greater availability, and fewer image artifacts.[4] Also, it is an easier procedure to do. SPECT provides more-than-adequate resolution for our applications.

CONCERN: Radiation exposure, especially in children.

RESPONSE: The average radiation exposure for one SPECT scan is 0.7 rem (similar to a nuclear bone scan or brain CAT scan) and is a safe procedure, according to the guidelines established by the

4. George, M. S. *Neuroactivation and Neuroimaging with SPECT.* (New York: Springer-Verlag, 1991).

American Academy of Neurology.[5] These other procedures are routinely ordered for many common medical conditions (i.e., bone fractures or head trauma), further suggesting that the levels of radiation exposure are generally acceptable in medical practice. Ineffective treatment of psychiatric illness has many more risks than the low levels of radiation associated with a SPECT scan.

CONCERN: What is normal?

RESPONSE: In the SPECT literature over the past twenty years, there have been more than forty-three studies looking at "normal" issues in more than 2,450 patients, including 150 children from birth on (see www.amenclinic.com for references). These do not include the thousands of control subjects used in studies of specific neurological and psychiatric conditions. Chiron et al. reported that at birth, cortical regional cerebral blood flow (rCBF) was lower than those for adults.[6] After birth, it increased by five or six years of age to values 50 percent to 85 percent higher than those for adults, thereafter decreasing to reach adult levels between fifteen and nineteen years. At the age of three, however, children had the same relative blood-flow patterns as adults. Other common findings in normal studies suggest that women have generally higher perfusion than men and that age, drug abuse, and smoking have a negative effect on rCBF.

CONCERN: Some physicians say, "I don't need a scan for diagnosis; I can tell clinically."

RESPONSE: Often, well-trained physicians can tell clinically. But that is not when you order a SPECT scan. You order scans when

---

5. Report of the Therapeutics and Technology Assessment Subcommittee of the American Academy of Neurology: Assessment of Brain SPECT 1996; 46: 278–285.
6. Chiron C., C. Raynaud, B. Maziere, M. Zilbovicius, L. Laflamme, M. C. Masure, O. Dulac, M. Bourguignon, A. Syrota. Changes in Regional Cerebral Blood Flow During Brain Maturation in Children and Adolescents. *Journal of Nuclear Medicine* 1992;33(5):696–703.

you are confused, the patient hasn't responded to your best treatment, or the patient's situation is complicated.

CONCERN: Lack of reproducibility.

RESPONSE: The paper by Javier Villanueva-Meyer, MD, et al. elegantly answers this question, showing that there is less than 3 percent variability in SPECT scans over time for the same activity.[7] Our own clinical experience, scanning people sequentially, and sometimes twelve years apart, is that SPECT patterns are the same unless you do something to change the brain. SPECT is a reproducible and reliable method for sequential evaluation.

Conclusion

At the Amen Clinics we feel that our experience with more than 35,000 brain SPECT scans over sixteen years guides us in being the best in the world for brain-SPECT imaging.

Common Questions About Brain-SPECT Imaging

Here are several common questions and answers about brain-SPECT imaging.

Will the SPECT study give me an accurate diagnosis? No. A SPECT study by itself will not give a diagnosis. SPECT studies help the clinician understand more about the specific function of your brain. Each person's brain is unique, which may lead to unique responses to medicine or therapy. Diagnoses about specific conditions are made through a combination of clinical history, personal interview, information from families, diagnostic checklists, SPECT studies, and other neuropsychological tests. No

7. Villanueva-Meyer, Javier M.D. et al. "Cerebral Blood Flow During a Mental Activation Task: Responses In Normal Subjects and in Early Alzheimer Disease Patients." Alasbimn Journal: 1(3): http://www.alasbimnjournal.cl/revistas/3/villanuevaa.htm.

study is "a doctor in a box" that can give accurate diagnoses on individual patients.

Why are SPECT studies ordered? Some of the common reasons include:

1. Evaluating memory problems, dementia, and distinguishing between different types of dementia and pseudodementia (depression that looks like dementia)
2. Evaluating seizure activity
3. Evaluating blood-vessel diseases, such as stroke
4. Evaluating the effects of mild, moderate, and severe head trauma
5. Suspicion of underlying organic brain condition, such as seizure activity contributing to behavioral disturbance, pre-natal trauma, or exposure to toxins
6. Evaluating atypical or unresponsive aggressive behavior
7. Determining the extent of brain impairment caused by the drug or alcohol abuse
8. Typing anxiety, depression, and attention deficit disorders when clinical presentation is not clear
9. Evaluating people who are atypical or resistant to treatment.

Do I need to be off medication before the study? This question must be answered individually between you and your doctor. In general, it is better to be off medications until they are out of your system, but this is not always practical or advisable. If the study is done while on medication, let the technician know so that when the physician reads the study, he will include that information in the interpretation of the scan. In general, we recommend patients try to be off stimulants at least four days before the first scan and remain off of them until after the second scan if one is ordered. It is generally not practical to stop medications such as Prozac because they last in the body for four to six weeks. Check with your specific doctor for recommendations.

What should I do the day of the scan? On the day of the scan, decrease or eliminate your caffeine intake and try to not take cold medication or aspirin (if you do, please write it down on the intake form). Eat as you normally would.

Are there any side effects or risks to the study? The study does not involve a dye and people do not have allergic reactions to the study. The possibility exists, although in a very small percentage of patients, of a mild rash, facial redness and edema, fever, and a transient increase in blood pressure. The amount of radiation exposure from one brain SPECT study is approximately the same as one abdominal X-ray.

How is the SPECT procedure done? The patient is placed in a quiet room and a small intravenous (IV) line is started. The patient remains quiet for approximately ten minutes with his or her eyes open to allow their mental state to equilibrate to the environment. The imaging agent is then injected through the IV. After another short period of time, the patient lays on a table and the SPECT camera rotates around his or her head (the patient does not go into a tube). The time on the table is approximately fifteen minutes. If a concentration study is ordered, the patient returns on another day.

Are there alternatives to having a SPECT study? In our opinion, SPECT is the most clinically useful study of brain function. There are other studies, such as electroencephalograms (EEGs), Positron Emission Tomography (PET) studies, and functional MRIs (fMRI). PET studies and fMRI are considerably more costly and they are performed mostly in research settings. EEGs, in our opinion, do not provide enough information about the deep structures of the brain to be as helpful as SPECT studies.

Does insurance cover the cost of SPECT studies? Reimbursement by insurance companies varies according to your plan. It is often a good idea to check with the insurance company ahead of time to see if a SPECT study is a covered benefit.

Is the use of brain-SPECT imaging accepted in the medical community? Brain-SPECT studies are widely recognized as an effective tool for evaluating brain function in seizures, strokes, dementia, and head trauma. There are literally hundreds of research articles on these topics. In our clinic, based on our experience for over a decade, we have developed this technology further to evaluate aggression and nonresponsive psychiatric conditions. Unfortunately, many physicians do not fully understand the application of SPECT imaging and may tell you that the technology is experimental, but more than a thousand physicians and mental health professionals in the United States have referred patients to us for scans.

Glossary

Acetylcholine (ACh)—a neurotransmitter involved with memory formation, mostly excitatory, that has been implicated in problems with muscles, Alzheimer's disease, and learning problems.

Amygdala—found on the front, inside aspect of the temporal lobes, part of the limbic or emotional system of the brain, is involved with tagging emotional valences to experiences or events.

Anterior cingulate gyrus—runs lengthwise through the frontal lobes, brain's gear shifter, helping with cognitive flexibility.

Antioxidants—help to prevent damage from free radical formation.

Axon—usually a long process that projects from the cell body to connect with other cells.

Basal ganglia—large structures deep in the brain involved with motor movements, anxiety, and pleasure.

Central nervous system (CNS)—composed of the spinal cord and parts of the brain, brain stem, thalamus, basal ganglia, cerebellum, and cerebral cortex.

Deep limbic system—deep structures in the brain that influence emotional responsiveness.

Dendrites—structures that branch out from the cell body and serve as the main receivers of signals from other nerve cells, functioning as the "antennae" of the neuron.

Dopamine (DA)—a neurotransmitter involved with attention, motor movements, and motivation, has been implicated in problems with Parkinson's disease, attention deficit disorder, addictions, depression, and schizophrenia.

Estrogen—a female hormone of reproduction.

fMRI—a brain scan that uses powerful magnets to look at brain blood flow and activity patterns.

Free radicals—oxygen combined with other molecules to generate highly toxic substances that must be neutralized by antioxidants, or they damage cells.

G-spot—a controversial area of the brain on the front wall of the vagina, underneath the clitoris, thought to be extremely sensitive to touch and involved in intense pleasure and orgasm.

Gamma-aminobutyric acid (GABA)—an inhibitory neurotransmitter involved with calming brain function, has been implicated in problems with seizures, bipolar disorder, anxiety, and pain.

Ginkgo biloba—an herb from the Chinese ginkgo tree that is known to improve circulation and blood flow and has been shown to be helpful in dementia.

Glutamate—excitatory (stimulating) neurotransmitter.

Hippocampus—part of the inside of the temporal lobes that facilitates memory function.

Magnetic resonance imaging (MRI)—a brain scan that uses powerful magnets to look at the physical structure of organs.

Myelin—the whitish protein covering of neurons.

Myelination—the act of laying down myelin onto neurons.

Nerve growth factors (NGF)—one of several growth factors in the brain that promote the regeneration of nerve cells after injury.

Neurogenesis—the growth of new neurons.

Neuron—synonymous with nerve cell.

Neurotransmitter—a chemical that is released from one neuron at the presynaptic nerve terminal (the end of an axon), across the synapse where they may be accepted by the next neuron (on the dendrites) at specialized sites called receptors. There are many different neurotransmitters, such as acetylcholine, serotonin, dopamine, and norepinephrine.

Norepinephrine (NE)—a neurotransmitter involved with mood, concentration, and motivation and thought to be associated with problems of attention, depression, and anxiety.

Occipital lobes—visual cortex in the back of the brain.

Oxytocin—a hormone involved with bonding.

Parietal lobes—top, back part of the brain involved with sensory processing, visual processing, seeing movement, and direction sense.

Positron emission tomography (PET)—a brain scan that uses isotopes to look at glucose metabolism and activity patterns in the brain.

Prefrontal cortex—front third of the brain, responsible for executive functions such as forethought and judgment.

Progesterone—a hormone of reproduction.

Serotonin (5-HT)—a neurotransmitter involved with mood, flexibility, and shifting attention, is often involved with problems of depression, obsessive compulsive disorder, eating disorders, sleep disturbances, and pain.

Single photon emission computed tomography (SPECT)—a brain scan that uses isotopes to look at blood flow and activity patterns in the brain.

Synapses—junctions formed between nerve cells where the presynaptic terminal of an axon comes into "contact" with the dendrite's postsynaptic membrane of another neuron. There are two types of synapses, electrical and chemical.

Synaptic plasticity—the ability of synapses to change to more efficiently signal other neurons.

Temporal lobes—underneath the temples and behind the eyes, large structures involved in memory, auditory processing, mood stability, and temper control.

Testosterone—a hormone of reproduction.

References and Further Reading

Abramov, L. A., "Sexual Life and Sexual Frigidity Among Women Developing Acute Myocardial Infarction," *Psychosomatic Medicine* 38 no. 6 (1976): 418–25.

Addis, M., and J. Mahalik, "Men, Masculinity, and the Contexts of Help Seeking," *American Psychologist* 58 no. 1 (2003): 5–14.

Aharon I., et al., "Beautiful Faces Have Variable Reward Value: fMRI and Behavioral Evidence," *Neuron* 32 (2001): 537–51.

Ainsworth, M.D.S., et al., *Matters of Attachment: Assessed in the Strange Situation and the Home.* Hillsdale, N.J.: Erlbaum, 1978.

Allen, J. S., J. Bruss, and H. Damasio, "The Aging Brain: The Cognitive Reserve Hypothesis and Hominid Evolution," *American Journal of Human Biology* 17 (2005): 673–89.

Angier, Natalie, *Women: An Intimate Geography* (New York: Anchor Books, 2000).

Arnow, B., et al., "Brain Activation and Sexual Arousal in Healthy, Heterosexual Males," *Brain* 125 (2002): 1014–23.

Aron, A., et al., "Reward, Motivation, and Emotion Systems Associated with Early-Stage Intense Romantic Love," *Journal of Neurophysiology* 94, 327–37.

Azari, N., and D. Birnbacher, "The Role of Cognition and Feeling in Religious Experience," *Zygon* 39 no. 4 (2004): 901–17.

Azari, N., et al., "Neural Correlates of Religious Experience," *European Journal of Neuroscience* 13, (2001): 1649–52.

Bagley, Christopher, and Pierre Tremblay, "Suicidal Behaviors in Homosexual and Bisexual Males," *Crisis* 18 no. 1 (1997): 24–34.

Bancroft, J., "The Endocrinology of Sexual Arousal," *Journal of Endocrinology* 186 (2005): 411–27.

Bartels, A., and S. Zeki, "The Neural Basis of Romantic Love," *Neuroreport* 11 (2000): 3829–34.

Bartels, A., and S. Zeki, "The Neural Correlates of Romantic Love," *NeuroImage* 21 (2004): 1155–66.

Beatson, J., and S. Taryan, "Predisposition to Depression: The Role of Attachment," *Australian and New Zealand Journal of Psychiatry* 37 (2003): 219–25.

Begley, S., and A. Underwood, "Religion and the Brain," *Newsweek* 137 no. 19 (2001).

Berns, G. S., et al., "Predictability Modulates Human Brain Response to Reward," *Journal of Neuroscience* 21 (2001): 2793–98.

Bonanno, G. "Loss, Trauma, and Human Resilience, Have We Underestimated the Human Capacity to Thrive After Extremely Aversive Events?" *American Psychologist* 59 no. 1 (2004): 20–28.

Booth, Alan, et al., "Testosterone and Men's Health," *Journal of Behavioral Medicine* 22 no. 1 (1999): 1–19.

Bowlby, J. "Loss: Sadness and Depression," *Attachment and Loss*, vol. III, (New York: Basic Books, 1980).

Brunckhorst, C. B., et al., "Stress, Depression and Cardiac Arrhythmias," *Ther Umsch*, 60 no. 11, 673–81.

Bullough, Vern L., *Science in the Bedroom: A History of Sex Research* (New York: Basic Books, 1994).

Burleson, Mary H., et al., "Heterosexual Activity and Cycle Length Variability: Effect of Gynecological Maturity," *Physiology & Behavior* 50 (1991): 863–66.

Catania, Joseph A., and Charles B. White, "Sexuality in an Aged Sample: Cognitive Determinants of Masturbation," *Archives of Sexual Behavior* 11 no. 3 (1982): 237–45.

Charnetski, Carl J., and Francis X. Brennan, *Feeling Good Is Good for You: How Pleasure Can Boost Your Immune System and Lengthen Your Life* (Emmaus: Rodale Press, 2001).

Chockalingham, A., et al., "Estimation of Subjective Stress in Acute Myocardial Infarction," *Journal of Postgraduate Medicine* 355 (2000): 1969–70.

Chockalingham, A., et al., "Estimation of Subjective Stress in Acute Myocardial Infarction," *Journal of Postgraduate Medicine* 49 (2003): 207–10.

Chuang, Y. C., et al., "Tooth Brushing with Ictal Orgasms," *Seizure* 13 no. 3 (2004): 179–82.

Clymer, Adam, "U.S. Revises Sex Information, and a Fight Goes On," *New York Times,* December 27, 2002, A17.

Coffey, C. E., et al., "Relation of Education to Brain Size in Normal Aging," *Neurology* 53 no. 1 (1999).

Coleman, Eli, "Masturbation as a Means of Achieving Sexual Health," *Journal of Psychology and Human Sexuality* 14 no. 2/3 (2002): 5–16.

Curtis, Margaret, "Paradise Found? Hot Flash on the G-Spot," *Mademoiselle,* January 1989: 64.

Cutler, Winnifred B., *Love Cycles: The Science of Intimacy* (New York: Villard Books, 1991).

Cutler, W. B., E. Friedmann, and N. L. McCoy, "Coitus and Menstruation in Perimenopausal Women," *Journal of Psychosomatic Obstetrical Gynaecology* 17 no. 3 (1996): 149–57.

Darling, Carol A., J. Kenneth Davidson, Sr., and Colleen Conway-Welch, "Female Ejaculation: Perceived Origins, the Grafenberg Spot/Area, and Sexual Responsiveness," *Archives of Sexual Behavior* 19 (1990): 29–47.

Davey Smith, George, et al., "Sex and Death: Are They Related? Findings from the Caerphilly Cohort Study," *British Medical Journal* 315 (1997): 1641–44.

Davies, Stephenie, et al., "Sexual Desire Discrepancies: Effects on Sexual and Relationship Satisfaction in Heterosexual Dating Couples," *Archives of Sexual Behavior* 28 no. 6 (1999): 553–67.

Davis, Peter, and Ray Lay-Yee, "Early Sex and Its Behavioral Consequences in New Zealand," *Journal of Sex Research* 36 no. 2 (1999): 135–44.

Dennerstein, L., E. Dudley, and H. Burger, "Are Changes in Sexual Functioning During Midlife Due to Aging or Menopause?" *Fertility and Sterility* 76 no. 3 (2001): 456–60.

Ebrahim, S., et al., "Sexual Intercourse and Risk of Ischaemic Stroke and Coronary Heart Disease: The Caerphilly Study," *Journal of Epidemiology and Community Health* 56 (2002): 99–102.

Ellison, Carol Rinkleib, *Women's Sexualities* (Oakland: New Harbinger Publications, 2000).

Esch, T., and G. B. Stefano, "Love Promotes Health," *Neuroendocrinology Letters* 26 no. 3 (2005): 264–67.

Eslick, G., M. Jones, and N. Talley, "Non-cardiac Chest Pain: Prevalence, Risk Factors, Impact and Consulting—a Population Based Study," *Ailmentary Pharmacology and Therapeutics* 17 (2003): 1115–24.

Evans, Randolph W., and James R. Couch, "Orgasm and Migraine," *Headache* 41 (2001): 512–14.

Farah, M. J., "Why Does the Somatosensory Homunculus Have Hands Next to Face and Feet Next to Genitals? A Hypothesis," *Neural Computation* 10 (1998): 1983–85.

Feldman, Henry A., et al., "Low Dehydroepiandrosterone Sulfate and Heart Disease in Middle-Aged Men: Cross-Sectional Results from the Massachusetts Male Aging Study," *Annals of Epidemiology* 8 no. 4 (1998): 217–28.

Ferretti, A., et al., "Dynamics of Male Sexual Arousal: Distinct Components of Brain Activation Revealed by fMRI," *Neuroimage* 26 (2005): 1086–96.

Fisher, Helen E. *The Sex Contract—The Evolution of Human Behavior* (New York: Quill, 1982).

Fisher, H., et al., "Defining the Brain Systems of Lust, Romantic Attraction, and Attachment," *Archives of Sexual Behavior* 31 no. 5 (2002): 413–19.

Fisher, H., A. Aron, and L. Brown, "Romantic Love: An fMRI Study of a Neural Mechanism for Mate Choice," *Journal of Comparative Neurology* 493 (2005): 58–62.

Fogari, Roberto, et al., "Sexual Activity and Plasma Testosterone Levels in Hypertensive Males," *American Journal of Hypertension* 15 no. 3 (2002): 217–21.

Francoeur, Robert T., *Becoming a Sexual Person* (New York: Macmillan, 1991).

Gallup, G., R. Burch, and S. Platek, "Does Semen Have Antidepressant Properties?" *Archives of Sexual Behavior* 31 no. 3 (2002): 289–93.

Gangestad, S. W., R. Thornhill, and C. E. Garver-Apgar, "Women's Sexual Interests Across the Ovulatory Cycle Depend on Primary Partner Developmental Instability," *Proceedings of the Royal Society of London*, Series B 272 (2005): 2023–27.

Georgiadis, J., and G. Holstege, "Human Brain Activation During Sexual Stimulation of the Penis," *Journal of Comparative Neurology* 493 (2005): 33–38.

Gillath, O., et al., "Attachment-Style Differences in the Ability to Suppress Negative Thoughts: Exploring the Neural Correlates," *Neuroimage* 28 (2005): 835–47.

Goldstein J. M., et al., "Normal Sexual Dimorphism of the Adult Human Brain Assessed by *in Vivo* Magnetic Resonance Imaging," *Cerebral Cortex* 11 no. 6 (2001): 490–97.

Greenstein, A., et al., "Sexual Dysfunction in Women Partners of Men with Erectile Dysfunction," *International Journal of Impotence Research* 18 (2005): 1–3.

Guarraci, F. A., and A. Benson, "Coffee, Tea and Me: Moderate Doses of Caffeine Affect Sexual Behavior in Female Rats," *Pharmacology, Biochemistry and Behavior* 82 (2005): 522–30.

Gundel, H., et al., "Functional Neuroanatomy of Grief: An fMRI Study," *American Journal of Psychiatry* 160 (2003): 1946–53.

Gurian, M., *Mothers, Sons and Lovers: How a Man's Relationship with His Mother Affects the Rest of His Life* (Boston: Shambhala, 1994).

———., *What Could He Be Thinking? How a Man's Mind Really Works* (New York: St. Martin's, 2003).

———., *The Wonder of Girls: Understanding the Hidden Nature of Our Daughters* (New York: Pocket Books, 2002).

Habel, U., et al., "Same or Different? Neural Correlates of Happy and Sad Mood in Healthy Males?" *Neuroimage* 26 (2005): 206–14.

Hamann, S., et al., "Men and Women Differ in Amygdala Response to Visual Sexual Stimuli," *Nature Neuroscience* 7 no. 4 (2004): 411–16.

Hansen, B. "Partial Epilepsy with 'Ecstatic Seizures,' " *Epilepsy and Behavior* 4 (2003): 667–73.

Hiroshi, Kojima, Shozo, "Neuroanatomical Correlates of the Assessment of Facial Attractiveness," *Neuroreport* 4 (1998): 753–57.

Hite, Shere, *The Hite Report: A Nationwide Study of Female Sexuality* (New York: Macmillan, 1976).

Holstege, G., et al., "Brain Activation During Human Male Ejaculation," *Journal of Neuroscience* 23 no. 27 (2003): 9185–93.

Hommer, D., "Male and Female Sensitivity to Alcohol-Induced Brain Damage," *Alcohol Research and Health* 27 no. 2 (2003): 181–85.

Hurlbert, David Farley, and Karen Elizabeth Whittaker, "The Role of Masturbation in Marital and Sexual Satisfaction: A Comparative Study of Female Masturbators and Nonmasturbators," *Journal of Sex Education & Therapy* 17 no. 4 (1991): 272–82.

Ilmberger, J., et al., "The Influence of Essential Oils on Attention. I: Alertness," *Chemical Senses* 26 (2001): 239–45.

Janszky, J., et al., "Orgasmic Aura-a Report of Seven Cases," *Seizure* 13 (2004): 441–44.

Joseph, R., "The Limbic System and the Soul: Evolution and the Neuroanatomy of Religious Experience," *Zygon* 36 no. 1 (2001): 105–36.

Kaplan, Helen Singer, "Desire—Why and How It Changes," *Redbook* 58

(October 1984). As cited in Komisaruk and Whipple, "The Suppression of Pain by Genital Stimulation in Females."

Keesling, Barbara, *Rx Sex: Making Love Is the Best Medicine* (Alameda: Hunter House, 2000).

Komisaruk, Barry R., and Beverly Whipple, "The Suppression of Pain by Genital Stimulation in Females," *Annual Review of Sex Research* (1995): 151–86.

Krantz, D. S., et al., "Triggers and Timing of Cardiac Events," *Cardiology Clinics* 14 no. 2 (1996): 271–87.

Kreuter, M., et al., "Sexual Adjustment and Its Predictors After Traumatic Brain Injury," *Brain Injury* 12 no. 5 (1998): 349–68.

Ladas, Alice Kahn, Beverly Whipple, and John D. Perry, *The G Spot and Other Recent Discoveries About Human Sexuality* (New York: Dell, 1983).

Laumann, Edward O., et al., *The Social Organization of Sexuality— Sexual Practice in the United States* (Chicago: University of Chicago, 1994).

Lê, M. G., et al., "Characteristics of Reproductive Life and Risk of Breast Cancer in a Case-Control Study of Young Nulliparous Women," *Journal of Clinical Epidemiology* 42 no. 12 (1989): 1227–33.

Lehne, G. K., "Brain Damage and Paraphila: Treated with Medroxyprogesterone Acetate," *Sexuality and Disability* 7 no. 3/4 (1986): 145–58.

Levesque, J., et al., "Neural Circuitry Underlying Voluntary Suppression of Sadness," *Bioligal Psychiatry* 53 (2003): 502–10.

Levin, Roy J., "The Physiology of Sexual Arousal in the Human Female: A Recreational and Progreational Synthesis," *Archives of Sexual Behavior* 31 no. 5 (2002): 405–11.

Liu, R. S. N., et al., "Association Between Brain Size and Abstinence from Alcohol," *The Lancet* 355 (2000): 1969–70.

Marcus, M., and R. Miller, "Sex Differences in Judgments of Physical Attractiveness: A Social Relations Analysis," *Personality and Social Psychology Bulletin* 29 no. 3 (2003): 325–35.

McKenna, K., "Lecture 3: The brain Is the Master Organ in Sexual Function: Central Nervous System Control of Male and Female Sexual Function," *International Journal of Impotence Research* 11 suppl. 1 (1999): S48–S55.

McKenna, K., "The Neurophysiology of Female Sexual Function," *World Journal of Urology* 20 (2002): 93–100.

Meaddough, Erika L., et al., "Sexual Activity, Orgasm and Tampon Use

Are Associated with a Decreased Risk for Endometriosis," *Gynecologic and Obstetric Investigation* 53 (2002): 163–69.

Meston, C., and P. Frohlich, "The Neurobiology of Sexual Function," *Archives of General Psychiatry* 57 (2000): 1012–30.

Mouras, H., et al., "Brain Processing of Visual Sexual Stimuli in Healthy Men: A Functional Magnetic Resonance Imaging Study," *Neuroimage* 2 (2003): 855–69.

Murrell, T. G. C. "The Potential for Oxytocin (OT) to Prevent Breast Cancer: A Hypothesis," *Breast Cancer Research and Treatment* 35 (1995): 225–29.

Najib, A., et al., "Regional Brain Activity in Women Grieving a Romantic Relationship Breakup," *American Journal of Psychiatry* 161 (2004): 2245–56.

Nakamura, Katsuki, et al., "Neuroanatomical Correlates of the Assessment of Facial Attractiveness," *Neuroreport* 9 no. 4 (1998): 753–57.

Newberg, A., et al., "The Measurement of Regional Cerebral Blood Flow During the Complex Cognitive Task of Meditation: A Preliminary SPECT Study," *Psychiatry Research: Neuroimaging Section* 106 (2001): 113–22.

Odent, Michel, *The Scientification of Love* (London: Free Association Books, 1999).

Ogden, Gina, "Spiritual Passion and Compassion in Late-Life Sexual Relationships," *Electronic Journal of Human Sexuality*, http://www.ejhs.org/volume4/Ogden.htm (August 14, 2001; accessed November 22, 2002).

Palmore, E., "Predictors of the Longevity Difference: A Twenty-Five Year Follow-Up," *The Gerontologist* 22 (1982): 513–18.

Park, K., et al., "A New Potential of Blood Oxygenation Level Dependent (BOLD) Functional fMRI for Evaluating Cerebral Centers of Penile Erection," *International Journal of Impotence Research* 13 (2001): 73–81.

Pease, Allan, and Beverly Pease, *Why Men Don't Listen and Women Can't Read Maps* (New York: Broadway Books, 2001).

Persson, G., "Five-year Mortality in a 70-Year-Old Urban Population in Relation to Psychiatric Diagnosis, Personality, Sexuality and Early Parental Death," *Acta Psychiatrica Scandinavica* 64 (1981): 244–53.

Peski-Oosterbaan, A., et al., "Noncardiac Chest Pain: Interest in a Medical Psychological Treatment," *Journal of Psychosomatic Research* 45 no. 5 (1998): 471–76.

Petridou, E., et al., "Endocrine Correlates of Male Breast Cancer Risk: A Case-Control Study in Athens, Greece," *British Journal of Cancer* 83 no. 9 (2000): 1234–37.

Puri, B., et al., "SPECT Neuroimaging in Schizophrenia with Religious Delusions," *International Journal of Psychophysiology* 40 (2001): 143–48.

Ramachandran, V. S., and W. Hirstein, "The Perception of Phantom Limbs," *Brain* 121 (1998): 1603–30.

Reinisch, June M., *The Kinsey Institute New Report on Sex* (New York: St. Martin's Press, 1990).

Reiss, Ira L., and Harriet M. Reiss, *An End to Shame: Shaping Our Next Sexual Revolution* (Buffalo: Prometheus Books, 1990).

Rozanski, A., J. Blumenthal, and J. Kaplan, "Impact of Psychological Factors on the Pathogenesis of Cardiovascular Disease and Implications for Therapy," *Circulation* 99, 1999: 2192–2217.

Satcher, David, *The Surgeon General's Call to Action to Promote Sexual Health and Responsible Sexual Behavior 2001* (Rockville: Office of the Surgeon General, 2001).

Saver, J., and J. Rabin, "The Neural Substrates of Religious Experience," *Journal of Neuropsychiatry* 9 no. 3 (1997): 498–510.

Savic, I., et al., "Smelling Odorous Sex Hormone-like Compounds Causes Sex Differentiated Hypothalamic Activations in Humans," *Neuron* 31 (2001): 661–68.

Sayle, A. E., et al., "Sexual Activity During Late Pregnancy and Risk of Preterm Delivery," *Obstetrics and Gynecology* 97 no. 2 (2001): 283–89.

Shapiro, D., "Effect of Chronic Low Back Pain on Sexuality," *Medical Aspects of Human Sexuality* 17 (1983): 241–45. As cited in Komisaruk and Whipple, "The Suppression of Pain by Genital Stimulation in Females."

Singh, Devendra, et al., "Frequency and Timing of Coital Orgasm in Women Desirous of Becoming Pregnant," *Archives of Sexual Behavior* 27 no. 1 (1998): 15–29.

Sinha, R., et al., "Neural Circuits Underlying Emotional Distress in Humans," *Annals of the New York Academy of Sciences* 1032 (2004): 254–57.

Smith, G. D., S. Frankel, and J. Yarnell, "Sex and Death: Are They Related? Findings from the Caerphilly Cohort Study," *British Medical Journal* 315 (1997): 1641–44.

Sprecher, Susan, "Sexual Satisfaction in Premarital Relationships: Associations with Satisfaction, Love, Commitment, and Stability," *Journal of Sex Research,* 39 no. 3 (2002): 190–96.

Steptoe, A., and D. L. Whitehead, "Depression, Stress and Coronary Heart Disease: The Need for More Complex Models," *Heart* 91 (2005): 419–20.

Stevenson, R. W. D. "Sexual Medicine: Why Psychiatrists Must Talk to Their Patients About Sex," *Canadian Journal of Psychiatry* 49 no. 10 (2004): 673–77.

Stoleru, S., et al., "Neuroanatomical Correlates of Visually Evokes Sexual Arousal in Human Males," *Archives of Sexual Behavior* 28 no. 1 (1999): 1–21.

Swazbo, P. A., "Counseling About Sexuality in the Older Person," *Clinics in Geriatric Medicine* 19 (2003): 595–604.

Tur-Kaspa, Ilan, et al., "How Often Should Infertile Men Have Intercourse to Achieve Conception?" *Fertility and Sterility*, 62 no. 2 (1994): 370–75.

van Lunsen, R. H. W., and E. Laan, "Genital Vascuslar Responsiveness and Sexual Feelings in Midlife Women: Psychophysiologic, Brain, and Genital Imaging Studies," *Menopause: The Journal of The North American Menopause Society* 11 no. 6 (2004): 741–48.

von Sydow, Kirsten, "Sexuality During Pregnancy and After Childbirth: A Metacontent Analysis of 59 Studies," *Journal of Psychosomatic Research* 47 no. 1 (1999): 27–49.

Walters, Andrew S., and Gail M. Williamson, "Sexual Satisfaction Predicts Quality of Life: A Study of Adult Amputees," *Sexuality and Disability*, 16 no. 2 (1998): 103–15.

Wang, H. L., and J. F. Keck, "Foot and Hand Massage as an Intervention for Postoperative Pain," *Pain Management Nursing* 5 no. 2 (2005): 59–65, from *International Journal of Neuroscience* 115 (2005): 1397–1413.

Warner, Pamela, and John Bancroft, "Mood, Sexuality, Oral Contraceptives and the Menstrual Cycle," *Journal of Psychosomatic Research* 32 no. 4/5 (1988): 417–27.

WAS—World Association for Sexology, "Declaration of Sexual Rights," *14th World Congress of Sexology*, http://www.worldsexology.org/english/about_sexualrights.html (August 26, 1994; accessed October 15, 2002).

Weeks, David J. "Sex for the Mature Adult: Health, Self-Esteem and Countering Ageist Stereotypes," *Sexual and Relationship Therapy* 17 no. 3 (2002): 231–40.

Weeks, David J., and Jamie James, *Superyoung: The Proven Way to Stay Young Forever* (London: Hodder and Stoughton, 1998).

———. *Secrets of the Superyoung* (New York: Berkley Books, 1999).

Whipple, Beverly, and Barry R. Komisaruk, "Elevation of Pain Threshold by Vaginal Stimulation in Women," *Pain* 21 (1985): 357–67.

Wilson, M., and M. Daly, "Do Pretty Women Inspire Men to Discount the Future?" *Proceedings of the Royal Society of London, Series B* suppl. 271 (2004): S177–S179.

Wittstein, I., et al., "Neurohumoral Features of Myocardial Stunning Due to Sudden Emotional Stress," *New England Journal of Medicine* 352 no. 6 (2005): 539–48.

Yang, J. C., "Functional Neuroanatomy in Depressed Patients with Sexual Dysfunction: Blood Oxygen Level Dependent Functional MR Imaging," *Korean Journal of Radiology* 5 no. 2 (2004): 87–95.

Yavascaoglu, I., et al., "Role of Ejaculation in the Treatment of Chronic Non-bacterial Prostatitis," *International Journal of Urology* 6 no. 3 (1999): 130–34.

Zamboni, Brian D., and Isiaah Crawford, "Using Masturbation in Sex Therapy: Relationships Between Masturbation, Sexual Desire, and Sexual Fantasy," *Journal of Psychology and Human Sexuality* 14 no. 2/3 (2002): 123–41.

Zaviacic, Milan, et al., "Female Urethral Expulsions Evoked by Local Digital Stimulation of the G-spot: Differences in the Response Patterns," *Journal of Sex Research* 24 (1988): 311–18.

Acknowledgments

I am grateful to many people who have been instrumental in this book, especially all of the patients and professionals who have taught me so much about how the brain relates to our sexuality. The staff at Amen Clinics, Inc. has been of help and tremendous support during this process, especially Krystle Johnson and Niccole Miller. Catherine Miller provided inspiration and helpful critiques for the book and served as an amazing muse during some challenging times in this process; I am deeply grateful for her love and support. Nancy Benzel, PA-C, also provided love and support in the process and gave input on the hormonal issues discussed in the book. And I have deep gratitude to Drs. Douglas Kahn and Curtis Rouanzoin for their technical advice, especially on the "Use Your Brain Before You Give Away Your Heart" chapter, and for their care and concern for my own heart. Earl Henslin, Sheri Gantman, Leonti Thompson, Emily McGrath, Mark Laaser, Larry Momaya, Aisa Greene, Chris Hanks, Mary Knight, Jeff Smith, Mark Kosins, Barbara Wilson, Nancy Erickson, Dennis Alters, Steven Rudolph, CarolAnne Stockton, Cynthia Graff, David Bennett, Rosemary Jackson, and Tana Gebelin were instrumental in discussing the concepts in the book and reviewing the manuscript, as were my friends Chris Amen and Lucinda Tilley.

I also wish to thank my wonderful, fun, and loving literary team at Harmony Books, especially Shaye Areheart and Julia Pastore. I am ever grateful to my literary agent, Faith Hamlin, who is a constant source of strength, support, love, honesty, and encouragement.

Index

negative relational statements,
38–39
and paraphilias, 136, 137
and PMS, 185
positive relational statements, 38
problems with, 39, 40
in relationships, 38–39
and sense of smell, 154
and sex addictions, 125
ways to calm high activity, 40
in women, 79, 81
Delayed gratification, 27
Dementia, 118, 129, 234, 243
Denial, 205–8
Depakote, 185, 186
Depo Provera, 138
Depression:
antidepressants, 17, 190–91
in brain-injured patients, 96
and deep limbic system, 37, 38
and epinephrine and
norepinephrine, 60
and left hemisphere, 76
as mood disorder, 187–88
orgasms to treat, 17–18
and substance abuse, 218
symptoms of, 188
in women, 79
Detachment, 52, 68–72
Deutsch, Donny, 83
DHEA, 15, 16, 21
Diagnoses, 240–41
Diaperism, 128, 130, 135
Diet. See Food and diet
Direction(s), 75, 79, 85
Disconnectedness, 199
Divorce, 64, 67, 80, 85, 92–93
Domestic violence, 213, 214
Dopamine:
deficiency in, 26, 37
definition of, 245
and oxytocin and vasopressin, 67
in oysters, 161
role in attraction, 51, 59, 115
role in infatuation, 60–61

in seminal fluid, 191
and serotonin, 62–63
and sexual novelty, 174, 175
and Wellbutrin, 190
Dostoyevsky, Fyodor, 143
Doughnuts, 154–55
Drugs, 112, 126, 216, 218, 234
Duke University, 11
Dulse, 161
Dyspareunia, 174

Ebrahim, Shah, 13
Ecstasy, 41, 141–42, 144
ED. See Erectile dysfunction
EEG. See Electroencephalogram
Effexor, 122
EGB 761 extract, 153
Eggs, 159–60
Ejaculation, 11, 15, 167–68
Elderly, 79
Electroencephalogram (EEG), 242
E-mail, 226
Embedding, 176, 177, 178, 179
Emory University, 53
Emotional trauma, 214–16
Emotions, 37, 45, 46, 61, 80, 86,
137
Employment, 111
Endorphins, 16, 17, 159, 162
Entanglement, 174
Epilepsy, 2, 138, 142, 143
Epinephrine, 59–60
Erectile dysfunction (ED), 3
Erections, 55, 56, 152
Eros (god), 150
Erotogenic zones, 166
Escitalopram oxalate. See Lexapro
Essential oils, 156
Estrogen(s):
and brain cells, 78
definition of, 246
in female body, 75
as major sex hormones, 55,
56–57
and pain of PMS, 16

Estrogen(s) (*cont.*)
  in seminal fluid, 191
  and sexual attraction, 54
  and sexual frequency, 15
Exercise:
  as activity for couples, 82
  to alleviate depression, 69
  as aphrodisiac, 162–63
  avoiding before bed, 220
  and blood flow to brain, 36
Exhibitionism, 127, 129
Extramarital affairs, 64
Eye contact, 172
Eye movement desensitization
  reprocessing, 195

Facial expressions, 80
Family, 113
Fear, 41, 175
Feelings, 40, 80
Feet, 163–65
Fertility, 14, 66, 157
Fetishes, 123, 127, 128–36, 163
Figs, 160
Finances, 110, 213
Fingertips, 163
Fish, 221, 222
Fisher, Helen, 50, 60, 190
Fish oil, 222
Fitness, 19–20
Floral essential oils, 156
Flowers, 170, 171, 177
Fluoxetine. *See* Prozac
Fluvoxamine, 122
fMRI. *See* Functional MRI
Folic acid, 221
Food and diet, 156–62, 221,
  222
Foreplay, 85
Fortune Telling ANTs (automatic
  negative thoughts), 195–96
Free radicals, 246
Freud, Sigmund, 127
Friendships, 79
Frontal lobes, 24, 25, 76

Frontal temporal lobe dementia
  (FTD), 118–19
Frotteurism, 127
FTD. *See* Frontal temporal lobe
  dementia
Fueloep-Miller, Rene, 143
Functional MRI, 175, 242, 246

GABA. *See* Gamma-aminobutyric
  acid
Gallup, Gordon, 18
Gambling, 123–24
Games, 205–6
Gamma-aminobutyric acid
  (GABA), 41, 45, 246
Garlic, 160, 161, 162
Gender, 9, 73–88
  See also Men; Women
Genetics, 53, 113
Gestalt, 76
Gifts, 177
Giles, Graham, 15
Ginkgo biloba, 153, 246
Ginseng, 152–53
Glutamate, 246
God, 141
Goldstein, Irwin, 173–74
Gossip, 85–86
Grafenberg, Ernest, 165–66
Grandparents, 80
Grapefruit, 162
Gray matter, 76
G-spot, 16, 165–68, 246
Gur, Ruben, 81
Gurian, Michael, 82

Hair-coloring chemicals, 218
Hanging, 129
Happiness, 20–21
Harvard University, 58
Hasselbeck, Elisabeth, 83
Hatha yoga, 145
Headache, 17, 41
Head trauma, 211–16, 232, 234,
  243

reasons for, 241
side effects of, 242
variability-of-technique issues, 237–38
Sperm, 147
Sperm whale, 152
Spirituality, 113
Sports, 86, 205
SSRIs. *See* Selective serotonin reuptake inhibitors
Stigmatophilia, 128
Strangulation, 128, 129
Stress, 11, 41, 110, 222
Stroke, 13, 76, 78, 243
Substance abuse. *See* Alcohol; Drugs
Suicide, 79
Superyoung (Weeks), 19
Surprises, 171
Survival, 9
Sweat glands, 58, 200
Symmetry, 53–54
Sympathetic nervous system, 156
Synapses, 247
Synaptic plasticity, 247

Talk, 81, 82
Tantra, 144–46, 147, 162
Taoism, 147
Taste, 81, 179
Tegretol, 185
Television viewing, 86
Temper, 45, 46, 119, 213, 214
Temporal lobe epilepsy (TLE), 143
Temporal lobes (TLs), 24, 44–49, 138
    Amen Clinic Brain System Checklist, 228–29
    definition of, 247
    functions, 47–48
    negative relational statements, 46–47
    and paraphilias, 137
    positive relational statements, 46
    problems with, 47–48

in relationships, 46–47, 99
and sleep, 220
ways to balance, 48
10 Second Kiss, The (Kreidman), 179
Testicles, 55
Testosterone:
    benefits of, 15, 22
    and commitment, 64, 67
    at conception, 74
    definition of, 247
    effect of ambrein on, 152
    and male areas of brain, 75
    in seminal fluid, 191
    and sexual attraction, 54–56, 57
    similarity to musk scent, 155
    way to gauge amount of, 84–85
TGF beta, 18
Theobromine, 159
There's Something About Mary (movie), 126
Thompson, Leonti, 99, 134–35
Thoughtfulness, 170–71
Thoughts:
    and brain as "sneaky organ," 116–17, 140
    and brain function, 223, 225
    role of basal ganglia, 40
    of women, 81–82, 84, 87
Tibetan Buddhism, 146–47
Tics, 99, 119–20
TLE. *See* Temporal lobe epilepsy
TLs. *See* Temporal lobes
Toad secretions, 151
Toho University (Japan), 155
Torii, Shizuo, 156
Touch, 65, 81, 163, 165, 178–79
Tourette's syndrome, 119–20, 122
Tourette Syndrome and Human Behavior (Comings), 120
Toxic exposure, 218–19, 221–22
Transvestic fetishism, 128
Trauma, 211–16, 232, 234, 243
Triatomids, 151
Trichotillomania, 122

DANIEL G. AMEN, MD

DANIEL G. AMEN, MD, is a clinical neuroscientist, a psychiatrist, an author, and the CEO and medical director of Amen Clinics, Inc., in Newport Beach and Fairfield, California; Tacoma, Washington; and Reston, Virginia. He is a Distinguished Fellow of the American Psychiatric Association and Assistant Clinical Professor of Psychiatry and Human Behavior at the University of California, Irvine School of Medicine. Dr. Amen lectures to thousands of mental health professionals, judges, and lay audiences each year. His clinics have the world's largest database of brain images related to behavior.

Dr. Amen did his psychiatric training at the Walter Reed Army Medical Center in Washington, D.C. He has won writing and research awards from the American Psychiatric Association, the United States Army, and the Baltimore-DC Institute for Psychoanalysis. Dr. Amen has been published around the world. He is the author of numerous professional and popular articles, twenty-two books, and a number of audio and video programs. His books have been translated into fourteen languages and include *Change Your Brain, Change Your Life*, a *New York Times* best seller; *Healing ADD, Healing the Hardware of the Soul, Healing Anxiety and Depression* (with Dr. Lisa Routh), *Preventing Alzheimer's* (with Dr. William R. Shankle), *Making a Good Brain Great*, which was chosen as one of the best books in 2005 by Amazon.com and which also won the prestigious Earphones Award for the audiobook, and *Magnificent Mind at Any Age*, a *New York Times* bestseller. Dr. Amen, together with the United Paramount Network and Leeza Gibbons, helped produce "The Truth About Drinking," a show on alcohol education for teenagers that won an Emmy Award for the Best Educational Television Show.

In addition, Dr. Amen has appeared on the *Today* show, *The View, The Early Show, 48 Hours,* CNN, HBO, and many other television and radio programs. Dr. Amen writes a monthly column for *Men's Health Magazine* titled "Head Check."

AMEN CLINICS, INC.

Amen Clinics, Inc. were established in 1989 by Daniel G. Amen, MD. They specialize in innovative diagnosis and treatment planning for a wide variety of behavioral, learning, and emotional problems for children, teenagers, and adults. The Clinics have an international reputation for evaluating brain-behavior problems, such as of attention deficit disorder (ADD), depression, anxiety, school failure, brain trauma, obsessive-compulsive disorders, aggressiveness, cognitive decline, and brain toxicity due to drugs or alcohol. Brain-SPECT imaging is performed in the Clinics. Amen Clinics, Inc. have the world's largest database of brain scans for behavioral problems in the world.

The Clinics welcome referrals from physicians, psychologists, social workers, marriage and family therapists, drug and alcohol counselors, and individual clients.

Amen Clinics, Inc., Newport Beach
4019 Westerly Place, Suite 100
Newport Beach, CA 92660
(949) 266-3700

Amen Clinics, Inc., Fairfield
350 Chadbourne Road
Fairfield, CA 94585
(707) 429-7181

Amen Clinics, Inc., DC
1875 Campus Commons Drive
Reston, VA 20191
(703) 860-5600

Amen Clinics, Inc., Northwest
3315 South 23rd Street
Tacoma, WA 98405
(253) 779-HOPE
www.amenclinic.com

AMENCLINIC.COM

Amenclinic.com is an educational, interactive brain website geared toward mental health and medical professionals, educators, students, and the general public. It contains a wealth of information to help you learn about our clinics and the brain. The site contains more than three hundred color brain-SPECT images, hundreds of scientific abstracts on brain-SPECT imaging for psychiatry, a brain puzzle, and much, much more.

VIEW MORE THAN THREE HUNDRED ASTONISHING COLOR 3-D BRAIN-SPECT IMAGES ON:

Aggression
Anxiety Disorders
Attention Deficit Disorder, including the six subtypes
Brain Trauma
Dementia and cognitive decline
Depression
Drug Abuse
Obsessive Compulsive Disorder
PMS
Stroke
Seizures

ALSO BY DANIEL G. AMEN, M.D.

Daniel Amen has news for you: Your brain is involved in everything you do. Learn to care for it properly, and you will be smarter, healthier, and happier in as little as fifteen days!

Making a Good Brain Great
The Amen Clinic Program for Achieving and Sustaining Optimal Mental Performance
$13.95 paper (Canada: $18.95)
978-1-4000-8209-4

Magnificent Mind at Any Age
Natural Ways to Unleash Your Brain's Maximum Potential
$24.99 hardcover (Canada: $27.95)
978-0-307-33909-6

Change Your Brain, Change Your Life
The Breakthrough Program for Conquering Anxiety, Depression, Obsessiveness, Anger, and Impulsiveness
$15.00 paper (Canada: $23.00)
978-0-8129-2998-0